CANNABIS

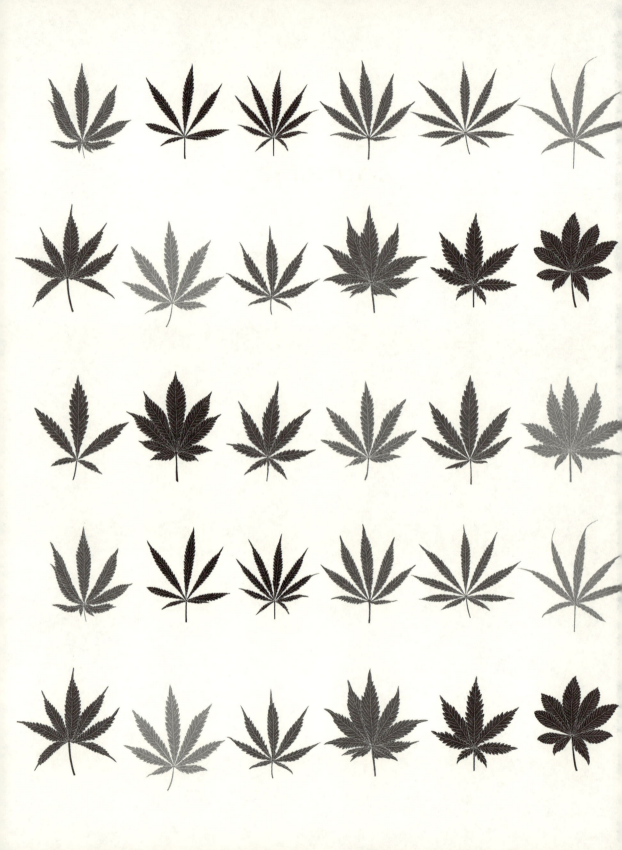

CANNABIS

A Natural History

ROB DESALLE

ILLUSTRATED BY PATRICIA J. WYNNE

Yale UNIVERSITY PRESS/NEW HAVEN & LONDON

Yale University Press books may be purchased in quantity for educational, business,
or promotional use. For information, please e-mail sales.press@yale.edu
(U.S. office) or sales@yaleup.co.uk (U.K. office).

Set in Adobe Text Pro type by Integrated Publishing Solutions.
Printed in the United States of America.

Library of Congress Control Number: 2025936356
ISBN 978-0-300-27094-5 (hardcover)

A catalogue record for this book is available from the British Library.

Authorized Representative in the EU: Easy Access System Europe, Mustamäe tee 50, 10621
Tallinn, Estonia, gpsr.requests@easproject.com

10 9 8 7 6 5 4 3 2 1

MIX
Paper | Supporting
responsible forestry
FSC® C008955
FSC
www.fsc.org

Dedicated to my brother Rollie: my closest friend, worthy rival, wise advisor, expert cannabis grower, and all-around wonderful guy. I will miss you.

Contents

Preface

Cannabis, the marijuana plant, offers an enthralling lens as a means to learn about the natural world. Cannabis is only one species of the billions of organisms that have ever lived on this planet. And like all of the billions of species that are alive or have lived (about 99.9 percent of all life on this planet has gone extinct in the past), cannabis is beautiful, whether you use it or not.

You might be wondering, Does this guy believe that every living thing is beautiful? Yes, I do, in some way or another, for the most part. The beauty is there mostly because of the sheer strangeness of all life, and more so because of the ways in which life has taken over this once sterile blob of chemicals and minerals we call Earth. Nothing that is living should be taken for granted! Even the lowliest of bacteria are amazing. The strangeness, improbability, vastness, and interconnectedness of Earth's ecological systems are what make them so beautiful, and also hard to explain.

How life on this planet got the way it is is one of those wonders of existence that is probably best pondered in one's more amiable states of mind. But we can find a lot of explanations logically. Science is the realm for doing this, and the challenge for scientists is to figure out the best explanations for the phenomena we see around us. It's tricky, because we are never going to know whether we are completely right about anything. We can convince ourselves we are right about something, but we are never going to be sure, because as the great philosopher of this problem, Karl Popper, realized, there is no final arbiter of truth in science; no one thing or person can tell us we are correct. But Popper did realize that we can determine when something is wrong or false—a scientific process he called falsification. Therefore, explaining cannabis is really an endeavor of identifying what is false or inconceivable about it, and using that as fodder for explanation. There are other ways to philosophize about this problem, but Popper's theory of empirical falsification has been quite helpful to me as a way of sifting through the information in the natural world.

I have found the biology of cannabis to be particularly astonishing. Its closest relatives (hops) and its place in the botanical tree of life might surprise you.

Though poorly understood until recently, it has a rich taxonomic history. Over a dozen different scientific names have been used for this plant species, which is in reality probably only one species.

Throughout this book I refer to cannabis by several names. For the most part I use only "cannabis" and "marijuana" as interchangeable terms for this amazing plant. Occasionally I slip into the vernacular and use "pot" or "grass," and when appropriate I will use the term "hemp." Hemp is an interesting name for cannabis, which usually refers to the plant's role as a fiber. It took on a different meaning in the United States when the national Agriculture Improvement Act of 2018 defined it as cannabis with less than 0.3 percent tetrahydrocannabinol (THC) content. I also alternate between using scientific names such as *Cannabis sativa* or the genus *Cannabis,* which I italicize. When I refer to the vernacular cannabis, I do not italicize or capitalize the word unless at the beginning of a sentence or in a title. The name marijuana is also vernacular, so I neither italicize nor capitalize it.

The sex life of cannabis lies at the heart of its utility, leading to the wholesale slaughter of male plants in marijuana fields. Males will pollinate the females, leading to a lot of seeds, which detracts from the production of the coveted chemicals tetrahydrocannabinolic acid (THCA) and cannabidiolic acid (CBDA). Although cannabis is often called a "weed," the variations it shows in outward physical appearance (phenotype) and its internal biochemistry are astonishing. For example, some marijuana plants are giants, growing to over ten feet tall, whereas others reach only a foot or two. As a consequence, there are over a thousand different strains. Cannabis plants make a thousand or more different chemicals that are stored in their tissues for a variety of evolutionary reasons, and these chemicals are the basis of marijuana's psychoactive and medicinal utility.

How did cannabis evolve the different pathways for the more than one thousand compounds it makes? How did it develop the sex system it uses to reproduce? Why are its floral structures the way they are? These questions require that we take a close look at its biology and genome (an organism's complete set of DNA), of which there are at least ten full genomes sequenced to date. An equally interesting question is, How did the many strains of this plant develop? Answering this question involves understanding the complicated genealogy of the strains. Because so much of the development of cannabis was accomplished surreptitiously, teasing apart its breeding history is more daunting than understanding wine grape genealogy, or even human genealogy.

Many readers will have their own anecdotes about their experience and encounters with cannabis, be they recreational or medicinal, exemplifying one of the major effects of THC on the brain. It prompts the release of dopamine (the feel-good receptor of the brain) and induces an overall feeling of euphoria. Because it influences the brain's receptors to be more active, it also heightens sensory perception and changes both the way you communicate and how you process incoming information. It also impairs your judgment.

Just as vivid as my pleasant memories of collegiate use of marijuana are my less mellow memories of the first time I saw the "This Is Your Brain on Drugs" televised public service announcement and heard Nancy Reagan's motto, "Just Say No." Both of these antidrug campaigns were representative of the backlash against marijuana use in the 1980s and the war on drugs of the 1980s and 1990s, much of which focused on cannabis use. These are important aspects of the maturation of marijuana use in modern society and its eventual partial legalization. They are also a reminder that there are negative connotations of marijuana use.

Not everyone is convinced that cannabis is nonaddictive. Some researchers have examined the role of marijuana smoking in the onset of schizophrenia, whereas others note the health problems associated with smoke inhalation and secondary smoke contact involved with marijuana use.

Although cannabis became a huge part of the counterculture of the 1960s and 1970s, it was already embedded in many other cultures. These cultural phenomena represent an important part of the plant's natural history. Its use in various rituals and religions is important for a full disclosure of its utility and history of usage.

If the patterns of use of cannabis by humans are any indication of its impact on culture and society, then the noble weed should be on everyone's mind. Over 60 percent of college students surveyed in 2020 said they have used marijuana at least once, with 16 percent of those imbibing at least once a day. Surprisingly, 60 percent is also the same frequency with which college students said they have used alcohol at least once. Younger people of high school age use marijuana and alcohol at half the rate of those who are of college age. Although use of cannabis has slightly dropped in adults, it is estimated that on average 20 million adults (around 10 percent of all adults) in the United States used marijuana in the past year. More surprisingly, it appears that over 60 percent of all Americans surveyed approve of some degree of legalization of marijuana. Indeed, over the past decade or so cannabis has been legalized in several states in the United States and in various other countries.

Legalization of cannabis has been slow and deliberate. As of this writing, twenty-three states and the District of Columbia allow fully legalized marijuana use and another twenty-three states have mixed-use laws, leaving only four states (Georgia, Kansas, Idaho, and Wyoming) where it remains fully illegal. It is difficult to generalize about the various degrees of legalization of cannabis use, because each state has its own laws, but it is clear that there is a move nationally toward legalization. The picture globally is a bit different, as in most countries marijuana use is still illegal. But many of the countries where it is illegal either have decriminalized it or do not enforce the laws they have on the books.

Recreational cannabis use is a cultural phenomenon that most users were forced to keep clandestine until recently. And its medicinal uses, although implied and in many ways anecdotal in the past, are only recently being quantified and seriously touted as legitimate medical phenomena.

Marijuana is also big business today. Several cannabis companies trade stock publicly, and it appears that the global COVID-19 pandemic has not been a hindrance to the growth and monetary value of these companies. It is not unusual to see IIPR, EGOV, SSPK, or GWRG (all ticker symbols for cannabis-based companies), to name a few, stream across the tickers on financial news channels.

As a full-time evolutionary biologist, part-time botanist, recreational user, and more than likely a future medicinal user, I find cannabis one of the most fascinating of all plants on our planet. I hope this book invites as much curiosity in you as it does in me—not only in cannabis but in science.

Acknowledgments

This book is the fourth in a series on the natural history of substances that we human beings use culturally, medically, and recreationally. In the spirit of the other books in this series, I want to first thank my mentor, friend, and coauthor Ian Tattersall. In 2007 Ian and I were reluctantly thrown together. Our collaborations began halfway through my tenure at the American Museum of Natural History and three-quarters of the way through his. We were asked to curate a new Hall of Human Origins, a task that we both believed was unnecessary; the Hall of Human Biology (the hall replaced by the Hall of Human Origins) looked fine to us and didn't seem to need to be redone. But if not for being thrown together on that project, I would have missed the most important opportunity of my career: to interact and learn from him (how to write, and how to enjoy wine), for which I am entirely grateful. I asked Ian to coauthor this volume, but he told me that he had no knowledge of cannabis other than it gave him insomnia. He intimated that he had put all of his energy into those other recreational substances that are more olfactory and gustatory—wine, beer, and spirits. I think, however, that he eschewed coauthoring again because he believed it was time for me to do something on my own. Thanks for the push, Ian. I only hope that this volume complements and achieves the same level of scholarliness as our other volumes. (He was mostly responsible for the scholarly merit of those volumes.)

I also want to thank several people for reading various early and probably painful drafts. These include Vivian Schwartz, who has read most of the other three volumes in this series for Ian and me. Her comments on readability and her ability to find topics that were missing from a comprehensive discussion are greatly appreciated. I thank Sara Oppenheim, Julia Zichello, Robbie Burk, and Douglas Daly for reading chapters concerning the neurobiology, biochemistry, and genomics of cannabis and the endocannabinoid system. All those chapters were greatly improved because of their comments. And a very special thank you to Dennis Stevenson and Chelsea Specht for reading the botany chapters and not laughing too hard at my lack of botanical knowledge. Chelsea was a shared student with Dennis and I, and it is a good thing for her that she learned to think like

Dennis and not like me. Her comments on the botany chapters reinforced how much she learned from Dennis and how little I know about plants.

I also want to thank the wonderful people at Yale University Press, especially Elizabeth Sylvia, who handled the initial processing of the manuscript; Margaret Otzel, who pored over the text for mistakes, corrections, and comments; and Jean Thomson Black, who meticulously edited the book. Jean especially has been a major influence on my career, always encouraging but always critical. I love working with her. Finally, I want to thank Patricia Wynne, whose wonderful art permeates this volume. Pat is perhaps the major reason that I write—just to see what she will draw to accentuate the writing. She is truly amazing at visualizing often dry information and enlivening the words on the pages of this book.

CANNABIS

1

Drunken Monkeys and Stoned Spiders

Questions such as, What is the history of cannabis? and Where did cannabis originate? can be answered at two levels—a deep historical level, replete with all kinds of ancient biological, geological, and chemical events that led to the genus *Cannabis,* and a shallow level where we delve into how humans influence the world in which we live, and how cannabis fits into that world. The former involves evolution, and the latter reflects culture. In examining the natural history of cannabis, we will consider both ways of explaining what we see—evolutionary and cultural.

Rules

Scientists engage in scientific and philosophical debates all the time. It is how we communicate how and what we have falsified, and what that might mean for explaining earthly (and extra-earthly) phenomena. One of my favorite colleagues, whom I have known for thirty-five years, once explained to me over a toke or two how the natural world works. It didn't matter that we had altered our consciousness; all of what we said must have made sense, as our hazy scientific discussions were always quite productive. He had three rules for how life on this planet works. As it turns out, all three of these rules contribute to the enhancement of survival or reproduction.

Why does reproduction matter? It is what drives all organisms—to make more of themselves, or their kind. The mechanisms of reproduction on this planet have built-in feedback loops that oblige organisms to make more of themselves, and the source of those feedback loops is related to the specific mechanisms of transferring information from one generation to the next. Feedback loops can be, as the famous biologist Richard Dawkins posited, reliant on selfishness. But the physical basis of this reliance involves the hereditary material of organisms (DNA, in most of them), and the way heredity works can also mean that randomness is involved.

Which brings us to Charles Darwin and what he got right. You might have thought Darwin discovered evolution and had gotten that right, but he didn't really. Evolution was known and written about well prior to Darwin's birth, most evidently to him by his grandfather Erasmus Darwin. While he did provide over-whelming evidence for evolution in nature (read chapter 4 in *On the Origin of Species* for his "long argument" for evolution), he did not discover it. What Darwin was first to get right (along with Alfred Wallace) was the principle of natural selection. However, just seeing evolution in nature (and it was undeniable to him and others at the time) doesn't mean you are finished. Such a phenomenon begs explanation, and Darwin's explanation was natural selection, for which he established that two major processes were necessary: differential survival, and differential reproduction.

Imagine that you are a cannabis plant. Because of changes in your genome (called mutations), you can produce novel chemicals that keep other organisms such as insects from eating you. If other cannabis plants growing around you don't have your novel mutation and can't make that chemical, then you will survive to reproduce while the others won't. You thus have a differential survival advantage over the other plants in your species, and your offspring will flourish over the others. Now imagine you are still a cannabis plant, but instead of a new chemical, a genomic change results in a flower with more seeds. When seed set occurs, you spew out more seeds than other plants around you, creating a higher probability that your seeds will be the ones to grow into adults in the next generation. That change implements a differential advantage in reproduction that enhances your chance of proliferating your genome into future generations.

My colleague's three rules are relevant to Darwin's way of thinking. Rule number one: *I run away from that.* It is always a good idea to keep away from things that can cause harm to an organism. If you don't make it to your reproductive age, you don't reproduce. Simple as that. Avoidance of death—either from predation or from eating something bad for you, or from being bitten by something poisonous, or a plethora of other fatal things—is essential to your making it to an age when you can reproduce. Of all the things that can harm some organisms, there is one category that is hard to run away from: microbes. Organisms can't see them with the naked eye (if they have eyes), and the ways in which plants "see" things don't usually pick up on microbes. Often microbes have initiated their detrimental effects before an organism knows it has been near them. No running away in that case. Microbes are an important aspect of the existence of cannabis, and the cannabis plant has taken some curious measures to deal with them.

Rule number two: *I eat or drink that.* Much of an organism's existence is tied up with finding nutrition to carry on its physiological functions, so that it can eventually reproduce. Plants do this quite differently than animals and fungi. Plants use water, carbon dioxide (CO_2), and light as sources of nourishment and energy during photosynthesis, whereas animals take their energy from ingesting foodstuffs. Some bacteria can photosynthesize too, but for the most part plants such as cannabis are the major photosynthesizers on our planet. As anyone who has grown this plant knows, it needs certain light conditions to grow, as well as water and rich soil. If these are unavailable or perturbed, growth will not result in a reproductively active adult.

Rule number three: *I have sex with that.* Most organisms on this planet ignore this rule. That is because most organisms are microbial and do not have sex, but rather reproduce by cloning themselves. But organisms such as cannabis and humans reproduce by having sex. There is a good reason for doing so (see chapter 2), and without sex or some way to reproduce, the first two rules would become moot and organisms would simply be dead ends.

Here is one good example of a reproductive dead end: All animals and plants have organelles in their cells, called mitochondria, which perform cellular respiration and produce energy for the cell. Because of the structure of the reproductive cells of plants and animals, female ova can reproduce their mitochondrial DNA and mitochondria efficiently and copiously. This means their mitochondrial DNA will be transferred to the next generation easily. But a male's sperm are not so good at making mitochondria, and hence not proficient at making mitochondrial DNA. They generate only a small number of mitochondria, and so transfer of male mitochondrial DNA during sex to the newly fertilized egg is terribly inefficient. This results in a lack of passage of male mitochondria, and male mitochondrial DNA, to the next generation. Male mitochondrial genomes simply don't make it to the next generation.

Rules Were Made to Be Broken

My colleague would argue that these three rules are enough to explain the diversity of life on this planet, and in many ways he would be largely on target. But as with any set of rules, they were made to be broken, although exceptions also "prove the rule." Plants such as cannabis are very effective both at breaking and proving these three rules. For instance, cannabis can't run (rule 1), but it has developed ways to move quite efficiently through wind dispersal of its seeds, with which it can spread over long distances. Even more important, cannabis has evolved a protective system that is common among plants that can't run from predators. Not needing to run generally involves two major strategies: make yourself distasteful, even poisonous, or make yourself difficult to swallow. Cannabis has taken the former strategy by evolving several noxious compounds that keep herbivores from dining too extensively on them. A good example of the latter strategy can be found in cacti, with their prickly leaves. Sometimes plants can also turn the tables on animal species by making their seeds resistant to animal digestion. Cannabis takes this strategy as well; after its flowers are eaten, the seeds are ingested and then pooped out somewhere by the animal, enabling the seeds to develop into new plants. This strategy is more or less a truly broken rule: *I let that eat me.*

Plants such as cannabis break rule 2 by definition because they literally do not "eat." They are what are called autotrophs or primary producers, as opposed to heterotrophs (most animals), fungi, and some bacteria that actually do eat autotrophs or other heterotrophs for nutrition. Plants use light, carbon dioxide, and water to extract nutrients for energy, and as a byproduct they release oxygen. But there are plants that are heterotrophs—restoring a broken rule. In fact, organisms break this rule so much that many novel names for their nutrition-getting strategies have been advanced: electrotrophs, electrolithoautotrophs, organotrophs, photoheterotrophs, chemolithoheterotrophs, and even radiotrophs (which use radiation as a source of nutrition).

Rule 3 is also messy for plants, which don't necessarily "see" their mates, so they can't really choose based on the outward appearance of a potential mate. Animals can in many instances pick their mates. Charles Darwin wrote about this at length when he described how some animals assess their potential mates on the basis of how they look, smell, or feel. For instance, male tephritid flies attract females by developing their eyes extended on stalks; the farther apart the eyes, the "sexier" the male fly appears to the female. Birds attract mates with coloration and behaviors; male bower birds, for example, build a love nest for courting females, attracting them with colored objects that are collected and placed strategically around it. While many plants have "sexy"-looking or -smelling flowers, the sexiness hasn't evolved because they want to be attractive to other plants; instead, they are trying to lure animals into physically interacting with them so that they can transport pollen or a fertilized seed elsewhere.

But some plants can be choosy about their mates. Many plants are hermaphroditic and can mate with themselves as well as with others, but this system often leads to inbreeding and deleterious genetic defects. Many plants have evolved ways to balance reproducing with themselves and finding mates that are not themselves. Cannabis plants are mostly unambiguously male or female, and not both (see chapter 5). But other mating systems have evolved in plants to prevent different species from reproducing with each other. These systems sometimes don't work well, as plants are notorious for crossing between broadly separated species and even genera.

Back to Rules

The three rules proposed by my colleague are similar to Ronald K. Siegel's "three drives" that he describes in his book *Intoxication:* the needs for food, water, and sex. But these needs involve as many broken rules as the three my colleague established. I have combined Siegel's water and food drives into one rule, and introduced a new rule: running away, or protecting oneself from predation. Siegel suggests that there is a fourth drive in many living organisms, especially in animals. This drive is to get intoxicated. This fourth drive is so important that it should be added to my colleague's initial three rules. So rule 4 is, *I get high on that.*

Siegel's rationale for this fourth drive is very straightforward. But it is hard to see how natural selection is involved, unless getting high contributes to the survival or reproductive capacity of organisms, or unless it is somehow hardwired into other processes or behaviors that have been affected by natural selection.

Happily, this fourth rule can be entirely explained by how neural function has evolved and works in animals. Pleasure seeking is a major aspect of neural function in animals with complex nervous systems, and ingesting cannabis, cocaine, opium, alcohol, or any other pleasure-providing drug is basically all about pleasure seeking. Pleasure seeking is important to the way animal brains work because it reinforces many fundamental behaviors. Pleasure is intimately connected to food, drink, and sex, and it is an important reinforcer of the behaviors associated with seeking food, drink, and sex. It is highly likely that, to reinforce pleasure seeking as a part of their essential and hardwired drives, many animals have evolved to seek any source of pleasure—including potentially lethal ones. Plants, bacteria, fungi, and animals without brains (yes, some animals such as jellyfish and sponges lack brains and even complex nervous systems) aren't susceptible to this fourth drive or rule, because they have no brains and hence no neurological pleasure centers.

Nature as Pusher

There are abundant examples in nature of animals getting high or drunk. Many such episodes are related to fermented sources of sugar, such as fruits that produce alcohol. Some of the more interesting examples of animals ingesting fermented food products can be found among primates. Katherine Amato and a score of her primatologist colleagues conducted a systematic study of primate drunkenness in 2021 and found that at least fifteen nonhuman primate species regularly imbibe alcoholic foodstuffs or liquids. In many cases, the phenomenon is widespread throughout wild primate species. The Brazilian brown howler monkey *Alouatta guariba* has seven independent populations from four different locations that were observed to ingest alcoholic foodstuffs. Within the higher primates, chimps (*Pan*), gorillas (*Gorilla*), and orangutans (*Pongo*) all partake of alcoholic fruit (table 1.1).

Fruit and primates go together like soup and sandwich. So much so that evolutionary physiologist Robert Dudley proposed what he calls the drunken monkey hypothesis, which suggests that an evolutionary relationship between arboreal fruit-eating primates and the fruits they ate led to the widespread consumption of alcohol by primates. Because primates needed to assess the ripeness of a fruit for its efficient use, and ripeness and fermentation are coupled, they became unavoidably connected to alcohol consumption. Dudley takes this idea

Table 1.1. ALCOHOLIC CONSUMPTION IN PRIMATES

Species	Group	Location	Groups	Fruits	Alcohol
Alouatta caraya	New World monkey	Brazil	1	1	*
Alouatta guariba clamitans	New World monkey	Brazil	7	7	m
Ateles geoffroyi	New World monkey	Mexico, Panama	5	7	m
Callithrix jacchus	New World monkey	Brazil	1	2	m
Cebus imitator	New World monkey	Costa Rica	6	4	m
Macaca athibetana	Old World monkey	China	1	2	*
Chlorocebus djamdjamensis	Old World monkey	Ethiopia	1	1	m
Macaca assamensis	Old World monkey	Thailand	1	2	l
Papio anubis	Old World monkey	Uganda	1	1	h
Pan troglodytes	Higher primate	Republic of Congo	1	3	*
Pan paniscus	Higher primate	Democratic Republic of the Congo	3	8	*
Gorilla gorilla	Higher primate	Republic of Congo	3	3	*
Pongo pygmaeus wurmbi	Higher primate	Indonesia	1	3	l
Eulemur fulvus	Lemur	Madagascar	1	2	h
Hapalemur meriodonalis	Lemur	Madagascar	1	2	l

Note: "Groups" refers to the number of social groups where alcohol consumption occurs. "Fruits" indicates the number of different species of fruit consumed. "Alcohol" refers to the alcohol content of the fruit, where * = not determined, l = low content (<0.1%), m = medium (0.1 to 2%), and h = high (>3%).

even further by suggesting that drunken monkeys' preference for fermenting fruit is why humans like to get drunk.

A wide range of organisms are pests on cannabis. Two researchers at the University of California at Berkeley, P. S. Messenger and A. R. Mostafa, listed 272 species of arthropods that live on cannabis plants. Unfortunately, their report has never been published (which either means it has never been peer reviewed or was actually rejected by peer review; I can't tell which from my searches), but it has hung on, zombielike, in the literature. Since access to this paper is limited, it is difficult to determine the validity of the 272 number, but some well-respected entomologists have cited it as evidence of the ubiquity of arthropod associations with cannabis. It is also difficult to determine which of these 272 species are pests on cannabis plants in a deep sense. Most of the non–South Asian pests are recent interlopers on marijuana plants and would thus be pests in a shallow sense. *Cannabis* originated and thrived for millions of years in South Asia and parts of southern China, so any arthropod–cannabis associations from that area would be better candidates for pests of cannabis in a deep historical sense. Fortunately, a two-page publication from 1976 by Suzanne W. T. Batra lists insect species from northern India that are associated with cannabis. The insect species found on cannabis plants in this area of Asia include flies, beetles, aphids, wasps, beetles, butterflies, grasshoppers, and even the noninsect spiders. These species are found as adults and as larvae on the plant, and some of them are characterized as eating cannabis tissue. Others congregate to predate on the insects that are eating parts of the plant. Basically, cannabis and insects go together like soup and sandwich too. They make such precise matches that different strains of marijuana plants from different geographical areas will have very specific insect species on them. This is rather like inferring someone's home from their clothing: lederhosen equals Germany, cowboy hat equals Texas. The entomological identification of pests found on cannabis plants can then be used to determine the area of origin of imported cannabis. Trevor K. Crosby and colleagues pointed out in the 1980s that the insects most associated with cannabis in its natural habitat can be used to determine exact locations of the plant's origin. This advance was important to drug law enforcers, who wanted to know where certain illicit shipments of marijuana were coming from.

John McPartland has summarized the extent of cannabis pests and their host plants (table 1.2). The majority of these pests are arthropods—insects specifically—and the only vertebrate target for damage is found for the plant's seeds. As far as entomologists can tell, the insects that live on cannabis plants do not become intoxicated. Many insects do not have cannabinoid receptors. These receptors bind cannabinoid chemicals that naturally occur in the body with exogenous cannabinoids such as THC and cannabidiol (CBD) that can be taken into the body. The lack of such receptors at least partly explains why insects can feed on cannabis and not be neurologically impacted. It appears that the plant's chemical defense, at least against insects, is not cannabinoid based. But because cannabis produces over four hundred compounds, some of them might be used by the plant for defense.

Table 1.2. COMMON CANNABIS PESTS

Seed and seedling	Flower and leaf, outdoors	Flower and leaf, indoors	Stalk and stem	Root
Cutworms	Hemp flea beetles	Spider mites	European corn borers	Hemp flea beetles
Birds		Aphids		White root grubs
	Hemp borers		Hemp borers	
Hemp flea beetles	Budworms	Whiteflies	Weevils	Root maggots
		Thrips		Termites and ants
Crickets	Leaf miners		Mordellid grubs	
		Leafhoppers		Fungus gnats
Slugs	Green stink bugs		Longhorn grubs	
				Wireworms
Rodents				

Source: McPartland (1996).

Some noninsect invertebrates do, however, have cannabinoid receptors. McPartland and colleagues have published studies in which they assay for binding of cannabinoids to the membrane tissues of organisms. This kind of experiment was used initially to determine that insects do not have cannabinoid receptors. Further assays showed that sea urchins, mollusks, leeches, sea squirts, earthworms, velvet worms (primitive arthropod relatives), rock lobsters, and nematodes all have cannabinoid receptor function. Some cnidarians (hydroids) but not all (like jellies and corals) have cannabinoid receptors, while sponges do not. Basically, it is hit or miss as to whether an invertebrate has cannabinoid receptors.

Spiders live on cannabis in northern India and are one of the many kinds of organisms found on the plant. But spiders more than likely do not have cannabinoid receptors. Take a look at the next spiderweb you come across; they are architectural wonders. This is why researchers often use the intricacy of a spider web as an indicator of neural togetherness. In the 1950s Peter Witt first used stoned spiders to characterize the possible neurological effects of drugs on spiderweb building. The idea is that if a spider is fed something that affects the weaving of a web, then the substance that was ingested more than likely affected the spider's nervous system. Even venerable institutions such as the National Aeronautics and Space Administration (NASA) have used this assay to study the effects of substances on spider nervous systems (fig. 1.1). Although it is obvious that certain substances (especially LSD and caffeine) affect the architecture of spider-

Figure 1.1. Drawings of webs woven by spiders under the influence of the indicated drugs. Adapted from Noever, Cronise, and Relwani (1995).

webs, it is not clear how. Since it is not clear that cannabinoid receptors exist in spiders, it cannot be determined whether the effect of marijuana comes from the cannabinoids in the plant or from some other compound.

Most vertebrates do, however, have cannabinoid receptors, although cannabis ingestion by vertebrate animals is somewhat rare. Rodents are known to nibble at cannabis plants. There is an amusing anecdotal report of a mouse passing out from eating marijuana leaves, replete with photographs of the mouse on its back, all four legs sticking straight in the air; the only thing missing to make it a perfect Tom and Jerry moment would be x's over its eyes. Although this instance is amusing, the only documentation for it are the photos—not valid documentation at all. There are also reports of moles eating the roots of marijuana plants. Larger animals such as cattle, goats, and sheep have been observed to nibble on cannabis plants, with some effects on their motor skills (their walking patterns become erratic).

Most of the concern about cannabis use by animals has revolved around domesticated animals such as pet cats or dogs, for which there are many publications. The National Capital Poison Control Center tracks intoxication events in pets. Ninety-six percent of reports concern dogs, and only 3 percent concern cats. (Without knowing some biology, this skew would seem rather odd, but it really isn't.)

One of the more common modes of ingestion is secondary intoxication via marijuana smoke. Another route to ingestion would be surreptitious ingestion of edible marijuana by one's pet. As anyone with a pet knows, they are nosy and continually hungry. They will sniff and eat anything they think they can get away with. But there is a difference between cats and dogs regarding marijuana ingestion. Most of the edible marijuana that people use is sweetened in some way. Cannabis-laced chocolate and cannabis-infused brownies, gummies, and marijuana butter are all common ways in which humans prep cannabis for eating. If you offer a cat a small bite of a candy bar, more than likely it will turn its nose up. Do the same with a dog, and you will be lucky to get away with your fingers. Cats cannot significantly taste sweet foods, whereas dogs do. Sugar and salt have specific structures that our taste buds can detect because of molecules we make called taste receptors (there are five kinds: salty, sweet, sour, bitter, and umami). Cats have an inactive form of the sweet receptor; no cats, even big ones such as lions and tigers, significantly taste sweetness. Dogs thus are attracted to sweetened cannabis products while cats are not, leading to more cases of cannabis poisoning in pet dogs.

Cats are known drug users, though, with catnip perhaps being their favorite. They can attain a level of euphoria from catnip similar to what humans attain from cannabis ingestion. The catnip plant *Nepeta cataria* is native to southern and eastern Europe, the Middle East, central Asia, and parts of China, overlapping partially with the native distribution of the *Cannabis* plant. Like cannabis, it has also been moved to most of the rest of the planet by humans, so it currently overlaps broadly with *Cannabis.* Although the psychoactivity of cannabis works through the cannabinoid receptors in the brain, catnip—or more properly the oil called nepetalactone that is produced by the plant—interacts with the brain's olfactory lobe. This lobe lies just inside the nasal passage and implements vertebrates' capacity to smell. Once the nepetalactone oil is detected by the olfactory system, signals are sent to the brain's hypothalamus and amygdala regions. The amygdala regulates emotional response in vertebrates such as cats, and the hypothalamus controls instinctual behavior. Stimulation of the amygdala by catnip might explain the euphoric behavior of cats on catnip. Stimulation of the hypothalamus might also explain the aggressive behavior of some cats when they overdose on catnip.

Human behavior on catnip was first described by Basil Jackson and Alan Reed, two medical doctors who encountered recreational human catnip use in the 1960s. Drug experimentation in humans was peaking at this point, and Jackson and Reed felt obliged to detail the phenomenon of catnip use by humans. Catnip prior to the 1960s had been used to counter amenorrhea (the inability of females to menstruate in cycle), chlorosis (a form of jaundice), and infant flatulent cholic (what it sounds like). It appears that the use of catnip for medicinal purposes ceased around the middle of the twentieth century. The four cases of recreational catnip use by humans described by Jackson and Reed indicated that one can get high from catnip smoking, but it takes a lot more catnip than cannabis to attain a similar high. In some of the cases there were adverse side effects, including headaches and malaise. Jackson and Reed's 1969 publication is rather famous for another reason, however.

Note that the labels of the two plants in their study (fig. 1.2) were reversed. This mislabeling in the prominent *Journal of the American Medical Association* (JAMA) led to a flood of amusing correction letters from readers. One response from John T. Petersik, MD, in particular piqued my interest, because in 1969 I was both a Boy Scout and a farm boy: "Any farm boy or Boy Scout knows that Figure 1 in the article by Basil Jackson, MD and Alan Reed, MD actually shows a catnip plant (*Nepeta cataria*) instead of *Cannabis sativa* as labeled." I did remember scouring our campsites for cannabis plants at the behest of our scoutmaster, although we never knew what he was doing with them. JAMA published twelve responses, almost all from medical doctors attempting to correct the mislabeling, and it listed the names of another ten or so doctors who noticed the mistake. Jokingly, the editors also reported that there were "856 irate and outraged cats who noted the unusual phenomenon." In a nod to the popularity of cannabis at the time, they also lamented, "Let the journal carelessly move the spleen into the thorax or lung into the pelvis, [and] a few mild restorative comments are received. But let us mix up catnip with cannabis and . . ." While the animal ingestion examples above are somewhat anecdotal, there are hundreds of scientific papers that have used animals as experimental subjects in cannabis studies. These studies mostly involved using mice, as they are easy to manipulate both physically and genetically.

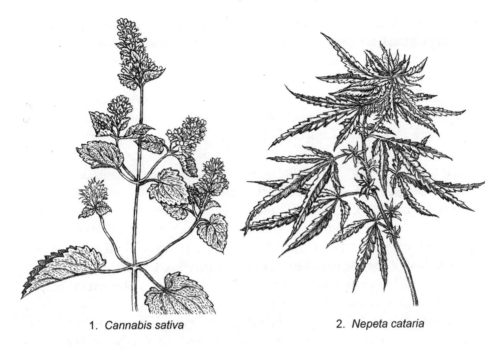

1. *Cannabis sativa*   2. *Nepeta cataria*

Figure 1.2. Drawings of *Cannabis sativa* and *Nepeta cataria* from Jackson and Reed (1969). The labels are as in the original publication.

No Stoned Monkeys

It is safe to say there is no stoned monkey hypothesis or even a stoned spider hypothesis for cannabis. Animals in general do not forage on cannabis plants, especially vertebrates and certainly not primates (except maybe for humans). Unlike alcohol, where there is a connection between feeding on fruit and ingestion of alcohol, there is no nutritive reason for primates to ingest cannabis. While cannabis seeds are a dietary source for some animals, most are turned off by it because the plant's generally nasty smell and bitter taste probably evolved as an antiherbivore tactic. The drunken monkey hypothesis is, moreover, a roundabout explanation for alcohol consumption in humans. According to the hypothesis, primates consume alcohol primarily because they want to ingest ripe fruit; the altered consciousness from alcohol comes through the back door. With drugs such as cannabis, the reliance on pleasure gained from altered consciousness has to come from somewhere else, because the drive to ingest something nutritious doesn't open the front door. Rather, as Siegel argues, there is a drive to intoxication that has evolved in organisms—which are often preadapted to using substances such as alcohol and cannabis.

In some ways humans were preadapted to ingesting marijuana. Alcohol in general is not a good thing to ingest for most animals. If there is too much of it in the body, the toxic effects of alcohol on the brain and other organs can be lethal. For some organisms, even a low level can be fatal. Animals have evolved enzyme systems to detoxify ingested alcohol that involve several genes called alcohol dehydrogenases (ADHs), numbered in the order of their discovery. These make proteins that can break down the alcohol. All animals have the major ADH molecule, and in humans about 10 percent of the total enzymes of the liver are these major ADH1 molecules. But it is another of the ADH molecules that initially encounters alcohol coming into the body as it is introduced in the tongue, esophagus, and stomach; this minor ADH (called ADH4) becomes an important enzyme for understanding alcohol ingestion in animals.

We are primates, and to understand the history of drunkenness we need to start with what our ancestors might have encountered. Several species of primates have been compared for the distribution of ethanol-targeting ADH4 molecules in their tissues. The species examined range from our rather distant relatives, the bush babies (galagos) of Africa and the aye-ayes of Madagascar; to the Old and New World monkeys; to our close chimpanzee relatives; and to our own species. The upshot of these comparisons is that there was a dramatic switch of the capacity of the ancestral ADH4 protein to process alcohol in the lineage that led to gorillas, chimps, and humans. This change led to a whopping forty-fold increase in the body's ability to metabolize ethanol, and thus to neutralize it. We share the powerful capacity of this ancestral enzyme with those other great apes.

In their book *Alcohol and Humans: A Long and Social Affair,* Kimberley J. Hockings and Robin Dunbar propose that African apes' new ability to deal with naturally occurring alcohol might actually have saved them from extinction. They argue that around 10 million years ago the apes (members of a formerly successful and diverse forest-living group that was by then in decline) were coming under increasing pressure from the flourishing cercopithecine monkeys. Specifically, those apes found themselves in competition with the monkeys for the ripe fruit that composed the core of their diets.

This explanation suggests that monkeys had the advantage in this competition, because they were able to process fruits at nearly *all* stages of ripeness. But there was one source of fruit for which the apes did not have to compete—the overripe fruit that had fallen to the forest floor, gently fermenting, as rotting fruit will do once invaded by wild yeasts. That fermentation produced alcohol—a new ingredient that, the argument goes, placed this particular resource off-limits to the monkeys, lacking as they did the enzyme needed to deal with it in any quantity. The newly alcohol-tolerant human-chimpanzee ancestors could therefore munch on those fermenting fruits as much as they liked. Chimps and humans were therefore preadapted to drinking alcohol, and lots more of it than your typical primate, prompting Robert Dudley to coin the term "drunken monkey." The question relative to marijuana then becomes, Were humans preadapted to be susceptible to cannabis? The answer is yes, but not in the same way the drunken monkeys prepared us for alcohol.

Because a broad range of animals have the capacity to react to the psychoactive compounds made by the cannabis plant, the evolutionary event leading to this ability must have happened long before our lineage diverged from other primate lineages. Cells in the neural tissue of animals are peppered on their surfaces with small molecules called receptors. The job of these receptors is to bind to specific small exogenous molecules that are signals for the brain to process all kinds of information. In cannabis, the small exogenous molecules in question are the cannabinoids previously referred to, and the small brain molecules are the cannabinoid receptors.

Cannabinoid receptor activity appears to have arisen during the early evolution of animals—or at least this is the most parsimonious explanation for their origin. Why? Because, as suggested earlier, they are found in almost all animal lineages that have brains. This means that most animals are programmed to react to cannabinoids, and that in turn most animals can get stoned on cannabinoids (unless by some chance they have lost the receptors, or the receptors have been altered in a way that renders them nonfunctional, like the sweet receptors of cats). But a crucial conundrum arises here. The genus *Cannabis* diverged from its closest living relative, the hops genus *Humulus,* only around 20 million years ago (mya), making that the earliest time that animals could psychoreact to marijuana cannabinoids. Animals arose at least twenty times that long ago (400 mya), which brings up two related questions. The possibility exists that other plants may have been making cannabinoids for much longer, meaning that it is possible that a common ancestor of hops and cannabis could have produced cannabinoids. A simpler explanation also exists: the lineage leading to *Cannabis* started making cannabinoids in the last 20 million years. So the question becomes, Are there other natural cannabinoids in other plants? And could other unrelated compounds in plants mimic some aspect of cannabinoids, and hence interact with cannabinoid receptors? Can molecules other than cannabinoids have cannabinoid receptors? The short answer to all these questions is "more than likely."

Regardless of the timing, we humans were preprogrammed to psychoreact to the cannabinoids produced by cannabis in the same way we were preadapted to alcohol via the drunken monkeys. Our species evolved with a brain fully capable of being impacted by cannabinoids. The question is whether this capacity was originally possessed by a stoned spider, a stoned sponge, a stoned starfish, or some more primitive stoned animal. Knowing the evolutionary history of cannabinoids can answer this question, as discussed in chapter 3.

2

Cannabis

A Short Cultural History

C annabis has played key roles in human history, human culture, and the development of human societies. Understanding its history and association with humans involves three distinct endeavors: one that follows the medicinal usage of the plant, one that follows its use as fiber, and one that follows its psychoactive use. Of course the first humans to use cannabis may have used it in any of several combinations of the three. These three potential uses require different methods for reconstructing past human behaviors within cultures, and how the kinds of plants and animals they interacted with influenced past behaviors. This complication limits the time frame we can use to around 20,000 years ago. There could have been a rich history of cannabis use before this time, but it is difficult to determine this without artifacts from archaeological sites of earlier ages (which currently do not generally exist). The documentation of early use of cannabis by humans thus has a ceiling that guides the arc of this chapter, and effectively what we know most about cannabis use by early Neolithic humans starts at around 12,000 years ago.

There are several ways in which scientists can implement the reconstruction of human cultural events. The first concerns the discovery of preserved material from paleontological or archaeological sites of known ages. This kind of evidence for cannabis use usually consists of ancient pollen, seeds, or burnt resin. Sometimes other artifacts, such as pipes or other ceremonial devices for cannabis smoking, have been found. Since the older estimates for the movement of humans into large social groups cluster around 12,000 years ago, the dating of the artifacts (both human-made and natural, like cannabis seeds) is quite precise. This precision is possible because 12,000 years is a nearly perfect age range for radiocarbon dating. Another approach is to glean information from ancient written texts; in other words, to note in which writings the cannabis plant is mentioned, and in what contexts. This approach is also quite accurate, as dates for most ancient texts are well established, but it is limited by the ages of the earliest discovered texts: the oldest known decipherable writing is Sumerian, only a little over 5,000 years old. Moreover, two things must fall into place if ancient writing is to tell us about cannabis use when the accounts were written. First, the writing must be decipherable; and second, it must contain something about cannabis use—and sadly, while many ancient scripts exist, the vast majority of them make no reference to cannabis. And there are some artifacts that contain what looks like writing but cannot be deciphered. For instance, while what is thought to be the oldest example of abstract representation can be found on the Blombos Cave plaques from South Africa originating 75,000 years ago, at this time we simply cannot interpret its meaning. Even if cannabis is mentioned on such an artifact, we will likely never know what its context was.

More recent writing on clay tablets and papyrus are the main sources of information for scholars attempting to reconstruct events in protohistoric times. These artifacts come from the Middle East, Egypt, and China, and are no more than 5,000 years old. Oral tradition, while not as precise as fossils and texts, offers another source of information in reconstructing past human cultural events. The oral tradition can be complicated and is, of course, only as good as the accuracy of the storytellers. Frequently evolutionary biologists use an approach called biogeography that can pin down the past movements of organisms. This approach assesses the current location of organisms (if fossils of an organism exist, all the better) to infer where the organism originated and where it moved to. With the development of molecular biological techniques that can extract and sequence DNA from ancient tissues (such as *Cannabis sativa* seeds), yet another approach has been added to the arsenal of scholars interested in reconstructing the past.

Fiber, Food (Fun), or Fitness First?

I examine several different kinds of historical events here: Where was the cannabis plant naturally found? Which cultures used cannabis? Which cultures used it as a fiber? Which cultures ingested it? Which cultures used it as a medicine? Each of these questions can also be accompanied by another "and when did . . ." question. First of all, any reader who wants the big picture fully fleshed out in all its glory should consult the wonderful treatise on cannabis origins, evolution, and movement by Robert C. Clarke and Mark D. Merlin, entitled *Cannabis: Evolution and Ethnobotany.* This amazing book chronicles in scholarly detail some of what I consider in this chapter. I will attempt to update and supplement the fascinating cannabis story that Clarke and Merlin started.

Whereas the story of cannabis and culture in this chapter starts about 12,000 years ago in Asia, we know that the plant has been around for a long time in both Asia and Europe. The archaeological evidence of humans associating with cannabis goes back to about 15,000 to 20,000 years ago. The question is, Were humans actually using it? Suffice it to say that the cannabis plant connects with human culture at almost exactly the same time that the first sedentary societies appeared. This means that cannabis had to be near those early sedentary human populations as they formed larger groups than the earlier hunter-gatherers had done. Since we will see that the original distribution of cannabis (its so-called area of endemism) was largely limited to the Tibetan Plateau (the general vicinity of Qinghai Lake), it makes sense to posit that early human groups living there inaugurated the cultivation and spread of cannabis as a crop plant. But did they?

To clearly establish the historical connection of cannabis to humans, there are checks to consider: Was cannabis present when human cultures began? If cannabis was present, was it being used as a fiber, being ingested (smoked or eaten), or being used medicinally?

Archaeological sites and core slices provide the two most accurate sources of information for exploring these checks. Because most plants spew their pollen and seeds everywhere they can, the gametes that don't participate in fertilization are sometimes readily preserved in sediment samples. Thus there is a natural record of plant pollen that can be obtained relatively easily. The record is extracted using core slice sampling. A researcher takes a cylindrical core sample from a site, the core is laid out and sliced across, and each slice is radiocarbon dated. After radiocarbon dating, the contents of each slice are examined under a microscope, and different kinds of pollen and seeds are identified. The shape and microstructures of both seeds and pollen are fairly diagnostic of the kind of plant that produced them. If a particular plant's seeds or pollen are found in the slice, then it is assumed that the plant itself was at the locality at the time the slice corresponds to. Upper and lower time ranges of a plant's presence can also be determined with this approach.

Although wild flax fibers have been found at an archaeological site at Dzudzuana Cave (dated at 30,000 years ago), located in the foothills of the Caucasus in what is now the country of Georgia, researchers had to wait for archaeological evidence from sites dated to 20,000 years later to find any similar use of cannabis. The first documented archaeological evidence of cannabis comes from Japan, where paleoecologist Makiko Kobayashi and colleagues examined ancient seeds from Jomon period archaeological sites. The Jomon period in Japan (2,500–14,000 years ago) can be defined by the different kinds of pottery the people of this culture made. However, the surviving seeds do not inevitably mean that the Jomon were the first cannabis users, for two reasons. First, other people may have been using it, but we have not yet found the archaeological evidence; and second, the Jomon may not have cultivated it consistently. Indeed, there is some debate whether the Jomon cultivated plants at all, or whether they simply practiced slash-and-burn subsistence. Recent evidence does suggest, however, that they were adept at cultivating plants that included chestnuts and hops (genus *Humulus,* the seeds of which can be confused with those of *Cannabis*). Other Jomon sites (Matsugasaki and Torihama: 5,000–6,000 years ago) suggest that cannabis was used as a fiber, so earlier use is not a far-fetched idea. Nonetheless, Akira Matsui and Masaaki Kanehara suggest that such plant husbandry was sporadic, and that widespread agricultural practice in Japan did not occur until the Yayoi period (1,500–2,500 years ago). Most researchers believe that cannabis was introduced to Japan and hence to the Jomon from China, and do not think that the Jomon were the first to use it.

Examination of time slices for cannabis distribution at archaeological sites can reveal whether humans were actually using cannabis. One study by Tengwen Long, William Hegman, and John McPartland reviewed the distribution of cannabis remains at the time before humans began living in large groups and domesticating plants and animals (about 12,000 years ago). The authors' analyses of European core samples are particularly enlightening in showing that attempts to pin down "when and where" using pollen are problematic for several reasons.

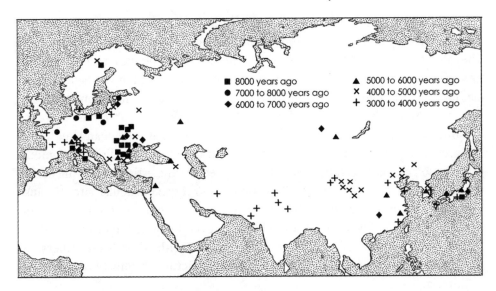

Figure 2.1. Map of the distribution of *Cannabis* in Asia and Europe at approximately 12,000 years ago from seed/pollen studies. Adapted from McPartland, Hegman, and Long (2019).

One of the major reasons is the close physical similarity of the pollen of the hops and cannabis plants because of their common ancestry. Many researchers simply refer to pollen that looks like hops or marijuana pollen as "CH pollen" (*Cannabis/Humulus*). Their work suggests that the first steppe communities that coalesced in the Tibetan Plateau may have been the first populations to associate with cannabis, and to implement its spread about 12,000 years ago. McPartland and his colleagues also indicate that cannabis was spread much earlier by natural agents such as wind or transport by animals. By the time human populations were coalescing into large sedentary social groups, cannabis can already be found in parts of Europe and Asia quite distant from the plant's area of endemism. McPartland and colleagues mapped the distribution of cannabis at 11,500 years ago in Europe and Asia (fig. 2.1), and it is more than likely that any human group in those occupied areas was using cannabis in some way. McPartland's group found cannabis pollen from Spain to Turkey as far back as 18,500 years ago. Was it moved there by humans, as we saw for the distribution at 4,000 to 5,000 years ago? We can't know, but we do know that the humans this far back in Europe were not farming it, and neither were humans elsewhere.

Ethan B. Russo, Michael P. Fleming, and Robert C. Clarke have performed meta-analyses of *Cannabis* specimens from cores collected at many Eurasian sites. They determined from the literature whether an archaeological site had pollen, or seeds, or paper products made from cannabis, or textiles or cordage made of cannabis fibers, or even the imprint of textiles or cordage. They came to strong conclusions, mostly in line with those of McPartland's group, about the presence of cannabis in China and surrounding areas, as well as at European sites. Since the archaeological sites in question have been dated using radiocarbon methods, Russo and his colleagues had relatively precise time estimates for the occurrence of cannabis at these sites. They focused on those particular sites because they were where nearly all of the early cannabis was found at the time of their meta-analysis. Tengwen Long has published several papers building on the initial meta-analyses. Since 2017 he had collaborated with several researchers, including McPartland, to develop a more complete picture of cannabis distribution in Eurasia.

The results of these analyses show that most of the dates for the occurrence of cannabis in Asia are less than 6,200 years ago and for Europe are less than 5,000 years ago (except for one questionable site in eastern Europe that is dated at 25,000 years ago and probably should be disregarded). One interesting result is that there is a marked increase in the number and locations of cannabis remains at about 5,000 years ago. Long and colleagues suggest that cannabis "might be associated with the establishment of a trans-Eurasian exchange/migration network through the steppe zone, influenced by the more intensive exploitation of cannabis achenes popular in Eastern Europe pastoralist communities." Indeed, a connection may have been made between cannabis cultivators in these regions. More precisely, Long and colleagues suggest that the Hexi Corridor served as a conduit for the transfer of materials and knowledge about hemp across Eurasia. The Hexi Corridor (also known as the Gansu Corridor) refers to a narrow stretch of relatively fertile plain in China west of the Yellow River. It was sandwiched between the desert terrain of Mongolia and the high Tibetan Plateau, and later became part of the famous Silk Road route.

Whereas pollen core data point to a broad distribution of the cannabis plant in Asia and Europe more than 15,000 years ago, the archaeological evidence for cultivation of cannabis is somewhat absent until 5,000 years ago. And here is where we separate the historical story of cannabis into three tales—Fiber, Food (Fun), and Fitness (Medicine).

Fiber

Clarke and Merlin devote an entire chapter of their magnum opus to cannabis as fiber and make it clear that the history of this use is complicated and tortuous. This convolution is caused by the fact that different people discovered the utility of hemp as a fiber independently and at different times. If the history of cannabis use had a simple vertical pattern of being handed from culture to culture in a sequential fashion, then the story would be quite linear and easier to understand. But while there is some verticality in the way cannabis spread as a fiber source, the independent discovery of its utility by many cultures complicates the story.

Humans have used plants for fiber for millennia (table 2.1). The bast, leaves, and seed hairs of plants make good raw material for fibers, and humans have been particularly adept at determining which plants could best be used as fibers. The stem of the cannabis plant has an interior part called the phloem, which is made of what are called the hurd (a wood-like interior part of the stem) and the bast (or its fibrous exterior; see chapter 4). The hurd is most commonly used in making paper and for insulation. The bast fibers of plants have been particularly useful in textile production because of their strength. After all, the bast and the hurd are what give the cannabis plant support to grow and stay upright. Before being used, the stems need to be retted to separate the bast from the hurd. This is done by laying the stems on the ground and rotating them for two weeks to a month. Over time the weathering on the ground separates the bast from the hurd, and the two components are then used differently. The retting process leaches out some of the cannabinoids in the cannabis plant, which can then be detected in core samples from archaeological sites. This factor has contributed to a better understanding of how various cultures have used the retting process on cannabis.

As the meta-analyses have concluded, central Asia was a hotbed of cannabis use, and most of the documented early use of hemp comes from archaeological sites in China. Clarke and Merlin deliver a tour-de-force analysis of cannabis as fiber, and focus most of their efforts on China for good reason—most of the active use of cannabis as fiber is from Chinese cultures starting at about 6,200 years ago, so much so that ancient China was known as "the land of mulberry and hemp." Much of the evidence of cannabis use for this area is historical, taken from preserved writings of this early period. The other evidence comes from archaeological sites. The earliest record of cannabis use as fiber in textiles is from Banpo, a Yangshao site that could be as old as 6,200 years. The evidence in question is a

Table 2.1. PLANTS USED AS SOURCE MATERIAL FIBERS BY HUMANS

Bast fibers		Leaf fibers		Seed hair fibers	
Flax*	*Linum usitatissimum*	Abaca	*Musa textilis*	Akund floss	*Calotropis procera* and *C. gigantea*
Hemp*	*Cannabis sativa*	Cantala	*Agave cantala*	Bagasse	*Saccharum officinarum*
Indian hemp	*Apocynum cannabinum*	Henequen	*Agave fourcroydes*	Bamboo	Various species
Jute	*Corchorus* species	Maguey	*Agave americana*	Bombax cotton	*Bombax* species
Tossa jute	*Corchus olitorius*	Mauritius hemp	*Furcraea foetida*	Coir	*Cocos nucifera*
White jute	*Corchus capsularis*	Phormium	*Phormium tenax*	Cotton*	*Gossypium* species
Kenaf*	*Hibiscus cannabinus*	Sisal	*Agave sisalana*	Floss-silk tree	*Ceiba speciosa*
Ramie*	*Boehmeria nivea*			Kapok	*Ceiba pentandra*
Roselle	*Hibiscus sabdariffa*			Milkweed floss	*Asclepias syriaca* and *A. incarnata*
Sunn	*Crotalaria juncea*				
Urena	*Urena lobata*				
Kudzu*	*Pueraria thumbergiana*				
Paper mulberry*	*Broussenetia papyrifera*				

Note: While hemp fibers have been used to make textiles for clothing, shoes, and sandals, for making paper, and for basket making, its most primitive use was most likely for cordage. An asterisk indicates that Clarke and Merlin consider these as used by Bronze Age humans.

textile imprint on pottery, the commonest category of artifact found at Chinese archaeological sites.

The earliest actual cordage and textiles from archaeological sites are from the Liangzhu culture in China, dated at 5,500 to 4,200 years ago. As the use of cannabis for textiles and cordage grew in China, the human spark got to work by

Ma – hemp

Hsi – male hemp

Chü – female hemp

Figure 2.2. Ancient Chinese symbols for hemp
(*ma*), male hemp (*hsi*) and female hemp (*chü*).

expanding its uses in art and other cultural endeavors. Archaeological sites have yielded artifacts demonstrating the use of hemp as burial cloths, canvas for art, and textiles for making banners. Burial practices were closely related to hemp use in ancient China, and even today hemp garments are worn as part of the grieving process. Silk was often embroidered into hemp backing to create cloth. Hemp cloth bags and enclosures were used to protect or preserve delicate items such as documents. Chinese lacquerware, famous for its beauty and durability, was also enhanced as far back as 3,000 years ago by hemp fiber layers applied during the lacquering process. Wigs were also made from hemp, but perhaps the most stunning use of hemp by the ancient Chinese was in warfare. Hemp was used first as tensile string for bows and longbows, and later as crossbow strings, as well as in the construction of catapults, as sails on warships, and for other military uses.

The written history of cannabis begins at about 4,700 years ago, in writings referring to the mythical Chinese Emperor Shennong (not to be confused with Wang Zhen's *Nong Shu* or *Book of Agriculture,* written 800 years ago, which also refers to cannabis extensively). In this ancient writing the emperor encouraged his people to grow and cultivate hemp. A multitude of subsequent texts, not only from China but all over the world, began referring to hemp as an agricultural crop.

To review these texts is beyond the scope of this book, but Clarke and Merlin offer a good summary. Hemp was also actively used in Korea, Japan, and Taiwan after 6,000 years ago. Its use in Taiwan and Korea was almost surely stimulated by its import from China. Although cannabis made it to Japan long before 6,000 years ago, the first Japanese uses of hemp as cordage and textile occurred somewhat later than in China, so it is difficult to tell whether the ancient Japanese concocted their own uses for the plant, or if all the concepts of the plant's usage came from China.

It appears likely that cannabis as fiber made its way to Mesopotamia and Egypt too, although the historical record is sparse, and it is assumed that its use as fiber in those regions was minimal at best. Use of hemp as fiber in Palestine and the surrounding areas inhabited by the Jewish tribes was at first considered lacking by scholars. But given that hemp fiber garments were part of the burial ritual of ancient Hebrews, this assumption is clearly wrong. In fact the word *cannabis* itself is probably derived from a Semitic language and was transferred first by the Scythians, who were well-known users of the plant.

Cannabis use as fiber moved into Europe at about 2,000 to 3,000 years ago. There is ample historical and archaeological evidence for its use in fibers in the Mediterranean, but there the focus was on cordage and sails for ships. Tracing the origin of words is often illustrative of how humans create and use things. We can trace the origin of the word *canvas* to its most common Mediterranean use in the past as sailcloth: *Canvas* comes from the Anglo-French word *canevaz,* which in turn derives from the Old French word *canevas,* which stems from the Latin word *cannapaceus* (which means "made from hemp"). The Latin word probably derives from the Greek word *kannabis,* which made its way to the Greeks from the Scythians. The Scythians probably adopted the original word from a Semitic language, according to John Fike from the words *kaneh* (meaning hemp) and *bosm* (meaning aromatic). The convoluted game of "telephone" that children play is the best way to describe this process of language transmutation. But it is amazing in this case how recognizable the words remained as languages developed.

Since cannabis started to be used as a fiber in Japan, its utility expanded, and the plant itself has been incorporated into Japanese culture. For instance, sumo wrestlers engage in a ritual called *dohyō-iri,* which involves the ceremonial parading of the *yokozuna* (the highest category of sumo wrestler) around the *dohyō* or ring prior to a match. Part of the *yokozuna*'s ceremonial garb is a huge rope, made from hemp, that adorns his waist. The rope is woven from hemp fibers by the members of the *yokozuna*'s stable and is called *tsuna,* which translates to "horizontal rope." Hemp has also become an important part of Zen Buddhist rituals and has been incorporated into Japanese fables.

Perhaps the cleverest Japanese parable concerning hemp is the well-known story of the ninja and the hemp plant. To increase the ninja's jumping skills, he is directed in this fable to plant cannabis seeds and, as soon as they sprout, to jump over the seedlings. Sounds easy, right? Not really, as the ninja must jump over the plants each day thereafter. Anyone who has watched a cannabis plant grow over several weeks knows that it will indeed grow quickly—and sometimes to a height of nine to twelve feet. A ninja with only moderate jumping skills at the beginning of the parable must develop astonishing athleticism by the end.

Hemp has also been cultivated and used as fiber in South Asia for some time, but as Clarke and Merlin point out, not as extensively as in East Asia. The oldest historical evidence for the use of cannabis as fiber in this region comes from the first of the Vedic Hindu scriptures (Kausitaki Brahmana of the Rig Veda). In this writing, from about 3,500 years ago, the sexual forms of cannabis are described. The Chinese also knew of the sexual differences of cannabis plants, as they had separate linguistic symbols for male and female plants (fig. 2.2). There is also mention of cannabis in the Mahābhārata Sanskrit text, which reports that Scythians imported hemp fiber in textiles about 2,800 years ago. The Scythians were a nomadic people also known as the Saka, who played a major role in the use of cannabis as a psychoactive substance. Overall, most East Asian peoples at that time were more interested in hemp as a medicinal, ritual, and psychoactive source than as a fiber source. Several authors point to the region's ready availability of other fibers such as ramie, jute, cotton, and flax, which in many ways made superior textiles.

Food (Feed and Fun)

The history of cannabis as food and as a psychoactive plant is much younger than its use as fiber. (Clarke and Merlin are the go-to text for in-depth analysis of this history.) How cannabis has served as food is a more straightforward story than how it has been used as a psychoactive or medicinal plant.

Many different tissues of the *Cannabis* plant can be used for nutrition, but some are more appropriate than others. Although some cultures have used the leaves in broth concoctions, it appears that most humans have settled on a single part of the plant as a source of nutrition, namely the seed. Humans have developed ways to ingest the seeds themselves, and to extract oil from them for ingestion.

Cannabis seeds have a high essential fatty acids content, as well as proteins, and can easily be ingested and processed by the human body. These are fatty acids such as Omega-3 and Omega-6 that are essential for normal human physiology, but our bodies do not synthesize them, so we need to get them from outside sources. Fish oils and oil seed plants such as flax are good sources of these important fatty acids, but hemp seeds also appear to have a suitable ratio of essential fatty acids to other chemicals for human consumption. Humans who lived during the hunter-gatherer period somehow knew that they could obtain these nutrients from plants and fish and included them as staples of their diets.

Evidence for hemp seeds as dietary items is older than the evidence for fiber use. The oldest known archaeological evidence for hemp seed usage as food comes from the Jomon era in Japan. Hemp seeds associated with the typical and diagnostic earthenware of the time were found at a site in the Boso Peninsula of central Japan, dated to 10,000 years ago, making them the oldest hemp fruit material on record so far. Other archaeological finds support the ancient use of hemp seeds as food in Japan, China, and Korea, as well as in what is now eastern Europe.

We are now ready to consider the earliest evidence for smoking cannabis. What kind of evidence do we need to determine the existence of smoking cannabis? Perhaps the most direct evidence would be finding burnt resinous remains of cannabis at an archaeological site. A second kind of information would be a written description of the practice of smoking. But the physical evidence for cannabis smoking trumps any written kind of evidence, although the younger instances of written evidence will become important for a full understanding of the history of smoking cannabis.

In 2019 archaeologist Meng Ren and six colleagues analyzed residue from nine wooden braziers found in western China at the Jirzankal Cemetery dating from 2,400 to 2,600 years ago. Apparently the braziers (fig. 2.3) were filled with stones, upon which plant material was placed and burned as an offering in funerary rites. It is a small step to imagine that the smoke from the offering was inhaled by the bereaved. That cannabis was associated with funerary practices in China is well established by the work of Hongen Jiang and colleagues, who discovered a grave at the Jiayi Cemetery of Turpan, northwestern China, where cannabis plants were laid over the corpse as a shroud. This site is dated a little before the Jirzankal Cemetery find (2,400 to 2,800 years ago), but because there was no evidence of burning or smoking, this site cannot be claimed as the first evidence of smoking

Figure 2.3. Drawings of two examples of wooden braziers found at the Jirzankal Cemetery in western China. Note the rocks in the example on the left, which were fitted into the hole in the brazier. Adapted from M. Ren et al. (2019).

marijuana. Ren and colleagues analyzed the burnt remains in the braziers using gas chromatography, which can determine whether the breakdown products of the cannabinoids in cannabis were present. Their analyses clearly showed the presence of cannabinol (CBN), cannabidiol (CBD), and cannabicyclol (CBL). Cannabinol is a breakdown product of both cannabidiol and tetrahydrocannabinol (THC), and its presence is convincing proof of the presence of both precursors in the resin. Because of the large amount of preserved residue on the rocks and in the braziers, Ren and colleagues concluded that the hemp plants burnt in the funeral rites must have had a substantial kick.

The first documented instance of smoking cannabis can thus be set at about 2,500 years ago. The history of psychoactive cannabis use gets richer as the writings referring to it become more recent. For instance, Herodotus, an ancient Greek scholar and historian who lived in Halicarnassus (now part of Turkey) about 2,300 years ago, wrote of cannabis use in the steppes near the Caspian Sea. In his *Histories,* he explicitly described cannabis ingestion by smoking among the people from the Caspian steppes. As for the Scythians, Herodotus described how, as part of their burial rituals, the bereaved would sit in small tents where the plants were burned on top of small stones: "After the burial, those engaged in it have to purify themselves, . . . they make a booth by fixing in the ground three sticks inclined towards one another, and stretching around them woolen felts, which they arrange so as to fit as close as possible: inside the booth a dish is placed upon the ground, into which they put a number of red-hot stones, and then add some hemp-seed."

Herodotus recounted the effect of this ritual further: the burning cannabis "gives out such a vapor as no Grecian vapor-bath can exceed; the Scyths, delighted, shout for joy, and this vapor serves them instead of a water-bath; for they never by any chance wash their bodies with water."

The smoking of cannabis hardly mellowed the Scythians; Herodotus's descriptions of this steppe people are quite brutal. According to him there was considerable decapitation, blood, and gore involved with this culture. For instance, the burial ritual of a Scythian king included the strangling of fifty of his young subjects, their disemboweling, and the filling of the body cavities with chaff, followed by sewing up the cavities. Simultaneously, fifty of the king's best horses were strangled and ceremonially placed upright by the use of stakes through their bodies; then the fifty chaff-filled human bodies were placed on the horses and also fixed upright by the use of stakes that ran parallel to their spines. The so-called riders were then placed in a circle around the king's tomb and left to rot.

While Herodotus might have exaggerated the grotesqueness of Scythian culture, he was probably fairly accurate about the cannabis smoking of the region's people. Ren and colleagues point out that artifacts from the frozen tombs of Pazyryk, found in the southern Altai Mountains in modern-day Russia, partially corroborate Herodotus's account of cannabis smoking. This site, dated at 2,500 years ago, held copper containers with stones and burnt and carbonized cannabis seeds inside them. While the Pazyryk tombs are far from the Caspian, the similarity of their artifacts to Herodotus's accounts is enough to lead to the conclusion that he got it right. But as Clarke and Merlin point out, the writings of Herodotus do not imply that the Scythians were smoking cannabis flowers the way we do today. The material being smoked was instead the seeds—although it does appear that this way of smoking cannabis was somewhat if not strongly intoxicating.

Psychoactive cannabis use moved quickly across Europe and into the Middle East. A 2020 analysis of resin on ceremonial altars at the Judahite Shrine of Arad demonstrates the occurrence of smoked cannabis at around 2,500 years in an area just west of the Dead Sea. This shrine holds two stone altars, upon both of which something was obviously burned as an offering, evidenced by dark resinous stains on top of the stones. Gas chromatography was used to determine what was in the burnt offerings. One altar was used for burning frankincense, a commonly used fuel for ritual burning, and the other altar was used for burning cannabis. The cannabis altar also contained traces of animal dung, which was probably mixed with the cannabis to make it burn better. It wasn't until 1,100 years ago that cannabis is thought to have entered Arab life via Persian and Iraqi sects on the border of the Islamic world that abutted the central steppes. Because the Koran did not forbid or even regulate the consumption of cannabis, it spread readily through Islam. Even when Fatimid ruler al-Hakim issued a law prohibiting the sale of alcohol, he too ignored regulating or banning cannabis in his caliphate.

The story of the origin of the word *hashish* begins in the Arab world in AD 1090 (about 900 years ago). Gabriel Nahas discussed this history in detail in 1982, starting with Hasan-i-Saban (more commonly known today as Hasan al-Sabbah). He was the leader of a group of fanatics called the Hashishiyahs. According to Marco Polo's account, Hasan-i-Saban relied on a secret potion that he fed to his young followers. Sex was also involved, as the main activity of the followers in Hasan-i-Saban's fortress was presumably "to make love to sensuous women." The sex and drugs put the young followers under an unbeatable spell that enabled Hasan-i-Saban to order them to kill his rivals. When the Crusaders heard of the existence of this fanatic sect, the word "Hashishiyahs" morphed into "assassins," and the word for killers of rivals was born. But where did the word "Hashishiyah" come from? The word *hashish* comes from the Arabic *ḥašīš,* which means "hay or dried herb." So while some might conclude that the secret potion that Hasan-i-Saban may have used to sway his followers was cannabis processed into the oily compressed substance we today call hashish, this might not be accurate. Nahas points out that while *ḥašīš* is indeed the origin of the word "Hashishiyah," it was conferred on Hasan-i-Saban and his followers as an insult. There was widespread contempt for the sect as a result of their excessive behaviors, and as Nahas writes, the derisive term for the group arose "to confer on the partisans of Hasan the low and disreputable character attributed by some scholars to hashish eaters rather than the actual devotion of Hasan's followers to the drug. It was a way to discredit them as well as hashish."

Despite the derogatory origin of the word *hashish,* this oily compressed form of cannabis became a prized drug in the Arab world, and indeed in more modern times a staple of cannabis culture worldwide. Cannabis as a psychoactive substance then spread rapidly into Europe and Africa and other parts of the world over the past thousand years.

Fitness (Physical Condition)

Evidence for the ancient medicinal use of cannabis is not as old as for fiber or for food (10,000 to 12,000 years ago), but is slightly older than for smoking (2,500

Table 2.2. MALADIES THAT WERE TREATABLE WITH
CANNABIS AS PER ANCIENT HUMANS

Medical condition	Modern evidence
Alleviation of memory loss	
Analgesic	✓
Anesthetic	
Antiasthmatic	✓
Antibiotic	✓
Anticonvulsive	✓
Antidepressive	
Antidiarrheal	✓
Antimigraine	
Antiparasitic	✓
Antirheumatic	✓
Appetite promoter	✓
Childbirth facilitator	
Hypnotic	✓
Reduction of fatigue	
Sedative	✓

Note: The check marks in the second column indicate maladies
that cannabis can treat according to modern research.
Source: Clarke and Merlin (2013).

years ago). The oldest evidence for the use of cannabis as medicine is relatively clear. The emperor Shennong (the same discussed in the fiber section of this chapter) clearly outlined the utility of cannabis in medicine, and he is believed to have lived 4,700 years ago. His writings are said to be collected in the *Pen-ts'ao ching* (*Divine Husbandman's Materia Medica*) from about 3,800 years ago. Considered the father of both Chinese agriculture and medicine, Shennong was believed to have had a transparent stomach in which he could observe the effects of ingesting various medicinal compounds. It is said that he would ingest tens of different plants at a time, and then sit and meditate upon their effects through the window to his stomach. While he recognized the medicinal importance of hemp, the *Pen-ts'ao* also pointed out that overuse would cause demons to appear.

Ancient cultures across Eurasia touted the utility of cannabis as a medicine. South Asia has a particularly rich tradition of using cannabis as a medicine, and such uses are well documented in ancient Indian writing. Other areas with a strong tradition of medicinal cannabis use include Southeast Asia, the Middle East, and Africa. The range of maladies that the ancients claimed to be treatable with cannabis was broad (table 2.2), and some of these have been substantiated by modern research. Oddly, evidence of early European use of cannabis as a medicine is lacking. In fact, some Roman medical texts point to the bad side effects of cannabis ingestion and avoid touting the plant as a medicine. As Europeans made deeper contact with India and China this reluctance to use cannabis as a medicine continued, and their traders remained more interested in cannabis as cordage than in any other use of the plant. The medicinal span of cannabis now, however, reaches across the globe.

3

Origins

How did cannabis get the way it is? How have living things in general become what they are today? The simple answer is, They evolved. But the peculiar aspect of cannabis's evolution is that it didn't become *Cannabis* until 20 million years ago. Before that it was a plant not too much unlike what it is now, but definitely not *Cannabis*. It was an ancestor of *Cannabis* and *Humulus* (hops), but it was neither *Cannabis* nor *Humulus*. This might sound like double talk, but it is indeed how *Cannabis* became what it is. It first had to have an ancestor with its closest relative (living or extinct): *Humulus*. And before that the *Humulus/Cannabis* ancestor had to have a common ancestor with several other plants that exist today—plants with names like *Pteroceltis tatarinowii* (blue sandalwood), *Celtis sinensis* (Chinese hackberry), *Morus alba* (white mulberry), *Ficus pumila* (climbing fig), and *Cecropia palmata* (snakewood tree). But let's start with the big daddy of them all—LUCA (the last universal common ancestor of all life)—and trace forward to *Cannabis*.

Let's Make a Deal

LUCA is the hypothesized last universal common ancestor of life on Earth. It was almost certainly a single-celled organism, somewhat like the bacteria we know today. There are several interesting ways LUCA could have arisen, but all we need to know for this discussion is that LUCA was real and existed 3.5 billion years ago. Among LUCA's direct descendants were the most important, dominant, and crucial life forms that ever flourished on this planet: cyanobacteria, a small, single-celled group of organisms that lived billions of years ago on a very hostile Earth without oxygen and with a highly chemically reduced atmosphere that some scientists describe as "Hadean" (after Hades). The cyanobacteria and their successors eked out their existence for a billion years in this harsh (to us, but perfectly fine to them) environment until about 2.45 billion years ago, when something transformative happened: the generation of significant amounts of oxygen. The cyanobacteria were not just passive bystanders taking advantage of the random emergence of oxygen on the planet; they were the agents making the most of it. Cyanobacteria were the organisms behind the great oxygenation event that made the planet able to support many novel kinds of life. Cyanobacteria, unlike us, gained their energy and made their living through processing sunlight. This process is called photosynthesis, and as its name implies they absorbed light (photons), water, and carbon dioxide, and converted these into nutrition and energy for the cells. That process gradually built up oxygen in the atmosphere.

The group of other single-celled organisms that lived along with cyanobacteria are known collectively as bacteria. They are one of the three major domains of organisms on the planet. The other two major lineages that evolved after the great oxygenation event gave rise to Archaea and Eukarya. For our *Cannabis* story, Eukarya is most important because it encompasses all plants, animals, yeast, and protists. These are grouped together because they all have a nucleus—a membrane-bounded area of a cell that encases the cell's genetic material. Eukaryotes probably arose about 1.8 to 2 billion years ago. (Archaea, which are single-celled without nuclei and mainly live in extreme environments, are mostly irrelevant to our story.)

Some strange events occurred as the eukaryotic lineage began to diverge into the familiar organisms we know today, including algae, amoebas, plants, animals, and fungi. Apparently, these early eukaryotes were hungry. Two of my colleagues at the American Museum of Natural History, Eunsoo Kim and Shin-chiro Muryama, suggest that the feeding frenzy of algal-like eukaryotes or cyanobacteria was not an unparalleled event. Rather, some of the ancestral algal cells were quite voracious and ate cyanobacteria of different kinds and at widely separated times in the past. Once ingested, the cyanobacteria stuck around inside the gobblers, like little aliens replicating and enjoying their inner existence inside the eukaryotic cells. In fact, the cyanobacteria that made it inside established a "deal" with the host cells that exists to this day, not only in algae but also in plants. The relationship was an "I scratch your back, you scratch mine" deal that ecologists call an endosymbiosis. The engulfed cyanobacteria became an essential part of the algal and plant cells, graduating from squatter status to full-blown organelles of these cells. The organelle they became is called a chloroplast; both plants and algae use their chloroplasts to harvest sunlight and produce food for their cells. Chloroplasts have a lot of chemical reactions churning away inside them, but the main one is the photosynthetic pathway.

The other single-celled organism that gave rise to another organelle called the mitochondrion was a proteobacteria-like ancestor that got gobbled up by the eukaryote ancestor. Which got gobbled first? More than likely the proteobacteria-like organisms, because all eukaryotes have mitochondria, but only plants and algae have chloroplasts. And the early eukaryotic cells were not only hungry but also in search of new ways to reproduce: they invented sex. Until eukaryotic cells invented sex, all inheritance of genomes, genes, traits, and so on was uniparental or clonal. Some eukaryotic organisms—usually single-celled ones—can reproduce clonally, but most of the multicellular ones have sex. There are of course exceptions to the rule, and plants have devised a number of ways to reproduce. The way ancient single-celled organisms generated variability was through mutation and clonal reproduction, creating lineages with the new mutations. Uniparental inheritance is quite a successful way to thrive, as evidenced by the millions if not billions of single-celled clonal species of bacteria, archaea, and some eukaryotes that reproduce clonally.

Sex

Sex changed it all for eukaryotes, though. As mentioned, some eukaryotic cells engulfed other cells (hence eukaryotes have mitochondria and chloroplasts), but the side effects of those cells getting engulfed also could do damaging things to the eukaryotic cells. When eukaryotic organisms went multicellular, the single-celled bacteria could start to use the multicellular organisms as hosts for other kinds of life. These parasites would rapidly evolve mechanisms to use their hosts that were almost always detrimental to the host. The eukaryotic organisms probably wouldn't have had a chance if not for sex. Many genes in organisms can have alternate forms called alleles. Sometimes one form for a gene (an allele) is better at the gene's job than another and hence has a selective advantage. Natural selection will favor that "better" allele. Different genes will interact with each other, too, to produce genetic combinations that are stronger in the face of natural selection. These other genes can have different alleles too. When that happens the fate of the alleles gets a bit complicated: instead of just one gene and two alleles, a eukaryotic organism could have two genes and four alleles. How does this all dovetail with sex?

Figure 3.1 (bottom) shows how variation is passed along in a uniparental or clonal organism. The organism starts out with two genes with one allele each—little a and little b. There are other alleles—big A and big B—that can arise by mutation and are better at surviving under natural selection. The object of the game is to get both big A and big B in the same lineage, starting with little a and little b. But mutations are rare, and the probability of getting two mutations for these two genes at the same time or even in steps is extremely low. Let's say a mutation occurs in two different lineages (which has a higher probability) about the same time to make two new genetic makeups—little a / big B (darkest gray) and big A / little b (medium gray). There is no way to get the two genes' favorable alleles together, other than to wait until one of the lineages has the second mutation (lightest gray). In the diagram the little a / big B lineage goes extinct, and the big A / little b lineage is the one that waits for the mutation of the little b to a big B. This takes time (lightest gray)—a lot of time!

Sex changes the timing. In the top diagram of figure 3.1, we start again with little a / little b, and the object is to get big A / big B for our two genes. Two mutations in the two different lineages occur to produce little a / big B and big A / little b combinations. But because the two lineages are "sexing it up," they can exchange alleles, which they will do as soon as contact between the two lineages occurs. This will happen almost immediately, and the big A / big B combination will be created and spread because it is the combination natural selection prefers. Sex offers a much quicker and a more efficient way to get those preferable combinations.

Time is of the essence too when natural selection is involved. A sort of arms

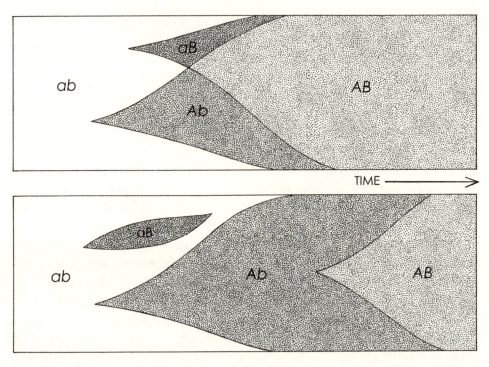

Figure 3.1. Diagram illustrating how sex (top diagram) might create novel combinations of geno-types more rapidly than asexuality (bottom diagram). The species starts with two normal alleles (a and b). Two advantageous alleles (A and B) occur at random through mutation in two different genomes. The two new advantageous alleles are recombined rapidly (top) through sex, much faster than with asexual organisms. The only way to get both A and B in an asexual system is to wait for the two mutations to occur in the same genome, an event much less probable than in a single genome with recombination. Adapted from Crow and Kimura (1965).

race develops between a parasite and a host. The parasite evolves a trick to take advantage of the host. If the host doesn't respond, it's bad news. But if the host evolves a defense quickly, then it will survive. But the parasite will escalate and evolve a new trick, and the host has to respond again. Sex enables the eukaryotic host to keep up with parasites in an evolutionary dance that biologists call the Red Queen hypothesis, referring to the Red Queen in *Through the Looking-Glass*, who had to run as fast as she could just to stay in the same place. That is basically what hosts need to do to coexist with parasites, and sex is a good way to run fast without stopping.

Biggies

We can boil the early evolution of plants down to some big events. While sadly, some of their beauty and splendor gets ignored by doing so, it makes it easier to understand where *Cannabis* came from. And every big event in this record is benchmarked by a major innovation or anatomical change in plants, which will eventually show up in *Cannabis*.

The first two major events led to the eukaryotic cells (getting the mitochondrion) and then to the plant cell ancestor (getting the chloroplast). The single-celled progenitors of plants were probably something like algae (but not really algae): the ancestor of plants and algae was something *different* from plants and from algae, but also something *like* plants and algae. Both lineages are descendants of this ancestor and diverged into several major sub-lineages—all the different colored algae (red, brown, and green in separate lineages with different restricted common ancestors) and plants with a single common ancestor.

The first big evolutionary event from this single-celled ancestor was to become multicellular. Liverworts, hornworts, mosses, and all other plants are living remnants of the ancestor formed by this multiculturalization event. These mat-like organisms have bizarre sex lives, with both sexual and asexual reproduction and some with male, female, and hermaphroditic sex parts. Their sex lives, however, are no more outlandish than most plants', so until we get to cannabis we will embrace chastity and not discuss sex. These organisms have very primitive plant characteristics and lack the vascular system of typical green plants, which is a giveaway of the next big event: vascularization.

Plant vascular systems are a little like our circulatory system, moving nutrients around the plant like our blood moves oxygen about our bodies. Food travels from the leaves to the roots, and water cycles from the roots to the leaves. Vascular tissue also provides woodiness, the "skeleton" of the plant, best seen in trees. The living descendants of this vascularized ancestor are lycophytes and ferns. Lycophytes, considered the more primitive of the two, do not have leaves as botanists define them. Ferns have a vascular system and have leaves where the plants do most of the photosynthesis, but it is thought that multiple independent evolutionary events led to fern leaves. If this is the case, then leaves probably evolved many times in the rest of the plant lineage. At this point our ancestral pre-*Cannabis* plant was vascularized, may have had leaves, and dispersed spores rather than seeds.

The next major evolutionary step, of course, was to produce seeds. The ancestor of all seed plants gave rise to the nonflowering gymnosperms (including cycads, pines, and ginkgoes), the angiosperms (including all flowering plants), and a strange plant group called Gnetales. The angiosperm ancestor had a vascular system, seeds, leaves, and the capacity to produce flowers. It also had a cotyledon, which became important for the next biggie.

The cotyledon, which means "seed leaf," is a structure that appears first after a seed has germinated. The seed can generate the cotyledon in two basic ways. It can form a single leaf extruding from the seed, or it can make two seed leaves. If it does the former it is called a monocotyledon (or monocot), and if it does the latter it is called a dicotyledon (or dicot). The ancestor that gave rise to all angiosperms had two seed leaves, as do the gymnosperms, so we need to add that to our list of traits in this ancestor: vascular system, seeds, leaves, flowers, and being dicotyledonous. But a funny thing happened on the way to corn, barley, other grasses, palms, and orchids, to name a few. The ancestor of these plants had the ability to develop a single seed leaf (was a monocot). *Cannabis,* on the other hand, is a dicot—a special kind of dicot called a core eudicot. Eudicot means "true" dicot, and core is used as a descriptor because the plants in the eudicot group are the more common examples of dicots, which number over 210,000 species. This means that to get to *Cannabis,* we need to follow our ancestors into eudicots, which gave rise to the majority of the beautiful flowering plants you are probably most familiar with (except for orchids).

Flowers

Perhaps the best way to examine flowers is to consider the most primitive flowering plant there is, and build on that flower's characteristics to arrive at a cannabis flower. What does "primitive" mean, though? It could mean oldest, but the oldest observed organism might not have primitive characteristics. Here is a good example from the animal world: There are five major kinds of animals—comb jellies, sponges, regular jellies and hydroids, bilateral animals (such as us), and a small nondescript animal called placozoa, which I will call a "pancake animal."

If one were just considering the anatomical characteristics of these five groups as an indicator of complexity, you would pick the placozoan as the most "primitive." It is clear from years of analysis, however, that these pancake animals are more closely related to jellies/hydroids and bilateral animals than to comb jellies and sponges; thus, the most simple does not equate with the most primitive. We have to be cautious when trying to interpret whether something living looks like a common ancestor or not. Even fossils can mislead us. As Daniele Silvestro and colleagues have pointed out, "a literal reading of the fossil record cannot be used to estimate realistically the time of origin of a group." Other considerations are involved, the most obvious being that the oldest fossil can only yield the youngest date of divergence. A fossil for a certain taxon may be 174 million years old, but the lineage to which that fossil belongs might be much older, simply because there might be another undiscovered fossil of the subject that is older. We just haven't found it yet.

Estimates of the age of flowering plants vary widely. The diversity of flowering plants exploded during the Cretaceous period (66 million to 145 million years ago) and became the dominant plant form in the Cenozoic (66 million years ago), so they have to be older than that. But because of the explosion of diversity during the Cretaceous, many botanists glommed on to that age range (66 to 145 million years ago) as the time of origin of flowers. There is one simple way to overturn this long-held idea of a Cretaceous origin of flowers, which is to find a flower fossil older than the Cretaceous. There is such a fossil called *Nanjinganthus dendrostyla,* discovered in China by Qiang Fu and colleagues in 2018 and dated to the Jurassic, approximately 174 million years ago. As the title of the Fu paper suggests, this was an "unexpected" find in the context of fossils. But if valid (and some paleontologists ignore this find), it would definitely push the origin of flowering plants back to at least 174 million years ago.

Another group of Chinese paleontologists claimed in a 2022 paper to have discovered Jurassic flower buds. But this find is also controversial, and most paleobotanists continue to assume that there are no credible flowering plant fossils earlier than the end of the Cretaceous. Nevertheless, it is clear that the fossil flowers discovered so far are at the earlier end of the Cretaceous, and that there was a flush of diversity around 130 to 100 million years ago.

There are several reasons why we don't see fossil flowers earlier than the early Cretaceous. First, the flower fossil record is sparse for that period, and given that flowers are notoriously fragile and don't fossilize well, they may just be missed. The second reason is that flowers then didn't look like flowers now, so while we don't see modern-looking flowers in the fossil record, we might not be looking for the right things. Paleobotanist Richard Bateman has called the elusive Jurassic flower a "snark" (referring to Lewis Carroll's mythical creature). He points to another recent fossil find by Xin Wang as critical. This fossil, *Lingyuananthus inexpectus,* was found in the Lower Cretaceous deposits at Yixian Formation in

Liaoning, China. It is unique because it looks like it has several derived characteristics that puzzle most paleobotanists. "Derived" means characteristics that are
found only in isolated modern plant forms. Because *Lingyuananthus* is not related to these modern plants, it suggests that Cretaceous plants' span of morphological characteristics is broader and stranger than previously thought.

An earlier age for flowers has not been unexpected, however, in the eyes of
molecular biologists, who use changes in the sequences of genes as a clock (much
like the ticking of radioactive material). We will return to this approach when we
revisit flowers in greater detail in chapters 5 and 6, but for now we can say that
flowers probably arose between 150 and 200 million years ago.

What do "derived" and "primitive" mean for paleobotanists? One definition
of primitive—"of, belonging to, or seeming to come from an early time in the
very ancient past," from the *Britannica Dictionary*—is probably what most people accept as the best definition. It's a good one, but primitive in the context of
how life on Earth has changed is better defined by Encylopedia.com as "relating
to, denoting, or preserving the character of an early stage in the evolutionary or
historical development of something." To understand this kind of primitive, we
need to examine the first branching flowering plant and its lineage. There is a
good candidate: a lineage with a single species called *Amborella*. This small plant's
lineage is what we call the "sister" to all other flowering plants. It's likely (but not
definite) that *Amborella* flowers resemble the first flower.

Amborella is a good first step, but what might the ancestral flower really
have looked like? One way to approach this problem with real data in a scientific
context is to undertake what is called ancestral reconstruction. This approach,
taken by evolutionary biologist Hervé Sauquet and colleagues, analyzes all the
anatomical information from living organisms and fossils and in a logical and al-
gorithmic way projects what characters might have existed in the common ancestor of all flowering plants.

This ancestral angiosperm flower would have been monoecious; it would
have both male and female parts. It would have had closed carpels (female reproductive parts), relatively long stamens (male reproductive parts), and undifferentiated perianths (nonreproductive part of the flower). This reconstruction
can even tell us how the nonreproductive parts (called tepals, sepals, and petals)
would have been arranged, whether they were differentiated sepals and petals or
simply tepals, and how many of these existed. The ancestral angiosperm flower
would have had ten or so whorled undifferentiated perianths, six or more stamens, and six carpels.

So far, the angiosperm ancestral flower thus described has an undifferentiated perianth with only a tepal. Most more derived flowers have differentiated perianths with both sepals and petals. Sepals are the parts of the floral envelope that are usually green and resemble tiny or reduced leaves, while petals are often colored and bizarrely shaped, immediately surrounding the reproductive parts of the flower. Tepals (note that tepal is a permutation of petal) are simply parts of the flower that are not petals or sepals. Sauquet and colleagues have also shown us that the ancestral flower of the group where *Cannabis* belongs looked like the ancestral Rosidae. This monoecious ancestral flower had reproductive parts that differentiated into five petals and five sepals. Eventually, *Cannabis* (along with *Humulus*) evolved a dioecious lifestyle; the genus differentiated their male and female reproductive parts into separate plants.

I may have gotten a little ahead of the story by describing the ancestral look of the cannabis flower, but we will return to cannabis flowers in detail in chapters 5 and 6. Just as the cotyledon evolved to give monocots and dicots, there were abundant evolutionary changes with angiosperms other than flowers. Let's take a step back and examine how angiosperms diverged.

A Rose Is a Rose Is a Rose; But Maybe Not

Along with dicotyledons, the ancestor of the eudicot group appears to have evolved some other characteristics that are distinctive to eudicots. Whereas monocots have long, narrow, parallel veined leaves, dicots have broad, reticulately veined leaves (fig. 3.2). Monocots' stems have their vascular system of bundles scattered, while dicots have vascular bundles arranged in a single ring. And the flowers of monocots display their parts in multiples of three, while dicots arrange their parts in multiples of four or five. So the ancestor of eudicots had a vascular system of a single ring, seeds, leaves that were broad and reticulately veined, flowers, seeds with two cotyledons, broad leaves, and flowers with parts in multiples of four or five.

The evolutionary relationships of eudicots are complex and hard to explain without mentioning many more species. Basically, the whole plant evolutionary business is so convoluted that Darwin himself called it an "abominable mystery." Suffice it to say that there are two major groups of eudicots, what botanists call supergroups: superasterids and superrosids. I must apologize to fans of buttercups (Ranunculales), lotuses (Proteales), and giant rhubarbs (Gunnerales). The apology also is extended to lovers of Trochodendrales (no common name, and only two known species exist) and Buxales (species with hard-to-pronounce com-

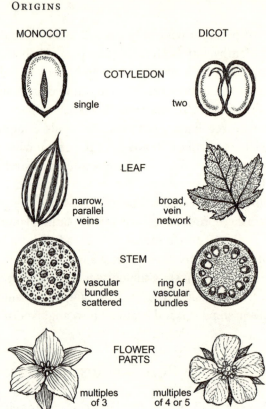

MONOCOT   DICOT

COTYLEDON

single   two

LEAF

narrow,   broad,
parallel   vein
veins   network

STEM

vascular   ring of
bundles   vascular
scattered   bundles

FLOWER
PARTS

Figure 3.2. Comparison of monocots and   multiples   multiples
dicots.   of 3   of 4 or 5

mon names, known to the Swedes as Buxbomsordningen). These five groups are core eudicots too, but I am going to ignore them and focus on the ancestor of the supergroup that contains *Cannabis*—the superrosids. (Perhaps I should also apologize to fans of sunflowers, daisies, holly, honeysuckle, ginseng, and the like—all of which are superasterids—as I will ignore their origin and ancestors too.)

Superrosids contain a group of plants that are responsible for another mind-altering substance: wine. In fact, grapes were the first group to diverge from the common ancestor of all superrosids. In other words, grapes can be considered phylogenetically primitive superrosids. Once the grapes peeled off the base of the superrosid tree of life, the ancestor of seventeen or so major kinds of rosids arose. This ancestor probably resembled the eudicot ancestor mentioned previously, but there are no distinctive anatomical characters that can describe the entire group (which would then logically be traits in the ancestor of the group). There are some tantalizing morphological traits that might have existed in the ancestor of these plants, including a structure called the hypanthium. This structure is also known as a floral cup or floral tube, and while it is present in a number of Rosales species, it changes too much over time and from species to species to be of use as a diagnostic for the whole group. When you use morphological traits to structure a taxonomic scheme, you want them to be stable and not to change as groups diverge. How then can we discriminate between the rosids and other plants if there are no anatomical traits we can use? The answer is that researchers have used genetic material to establish these relationships.

Of the seventeen major superrosid groups, *Cannabis* occupies one called the Rosales. Many of the plant names I have used are formal taxonomic epithets, meaning that they have been studied by taxonomists and validated by them based on a set of rigorous rules; but others are not so formal and accepted by botanists. An official procedure for naming plants includes following the rules determined by what was once the International Botanical Code of Nomenclature (IBCN) and currently is called the International Code of Nomenclature for algae, fungi, and plants (ICN or ICNafp). This code and the rules it advances are used by the International Association of Plant Taxonomists (IAPT) to develop valid names for plants that have been studied and for those that will be discovered in the future. (Animals face their own taxonomic problems and are governed by a zoological code as rigorous as that for plants.)

My descriptions of *Cannabis* ancestors so far have not been entirely kosher with respect to the ICNafp. It is important for a couple of reasons that we follow the ICNafp rules when we start to get specific about the relationships and names of things. Rosales are what biologists would call an order, and from here on I will be more mindful of the ICNafp rules. If you remember the hierarchy of taxonomic names (kingdom, phylum, class, order, family, genus, species) and the mnemonic that I use for it, "**K**ing **P**hilip **C**ame **O**ver **F**rom **G**reat **S**pain," you may have realized that I have so far ignored the naming conventions of the higher groups King, Philip, and Came (kingdom, phylum, and class). This is because plants sometimes don't follow these naming conventions at the level of King, Philip, and Came. I will now be more mindful of the formalities for "Over From Great Spain" (order, family, genus, and species).

Rosales (an O) is divided into eleven families (F's), of which *Cannabis* (a G) belongs to one called the Cannabaceae (one of the eleven F's). (Just as an aside, Rosaceae is another Rosales family that contains the roses, mulberries, figs, nettles, elms, and buckthorns.) The Cannabaceae is a small family subdivided into eleven genera (G's) and has about 100 species (S's). The ten genera other than

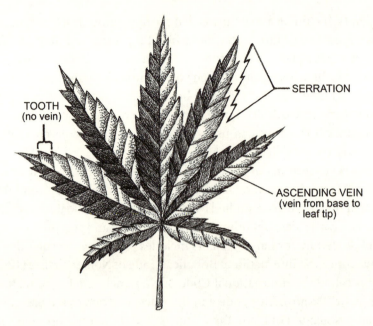

TOOTH
(no vein)

SERRATION

ASCENDING VEIN
(vein from base to
leaf tip)

Figure 3.3. Drawing of a typical leaf, showing the palmately compound leaf, the ascending veins, serrated leaves, and veinless teeth of the ancestor of Cannabaceae.

Cannabis include hops (*Humulus*), hackberries (*Celtis*), thorny elms (*Chaetachme*), pigeonwood (*Trema*), blue sandalwood (*Pteroceltis*), and five other small genera with small numbers of species in each.

There are some distinctive anatomical changes that existed in the common ancestor of all Cannabaceae. As anyone hiking who comes upon a wild cannabis plant knows, their leaves are quite distinctive. They have the highly distinctive palmately compound leaf, which has become the logo for many commercial outfits. They also have what are called "two-ranked, serrate leaves with ascending veins that do not proceed straight into the teeth." There are many ways the leaves of a plant can be arranged on the stem. A two-ranked leaf refers to opposite or alternate leaf arrangement, where the leaves are arranged in two vertical columns on opposite sides of the stem.

Two of the more commercially important genera in the Cannabaceae family—*Humulus* and *Cannabis*—have opposite and spiral leaves, respectively, but both are two-ranked, as are the other plants in the family. But I would guess that isn't the trait most easily recognized in the wild. The palmate leaves and their serration are what jumps out (fig. 3.3).

Figure 3.3 illustrates the technical jargon of the defining aspects of Cannabaceae leaves. But it's not as clean a story as one would hope; both *Humulus* and *Cannabis* genera have secondary veins running into the edge of the teeth and thus have veins within the teeth. Furthermore, the common ancestor of Cannabaceae most likely had fruits that were rounded drupes, which are simple fleshy berry-like structures with a single seed. But anyone who has been "stuck with seeds and stems again" knows that *Cannabis* (and *Humulus* too) have hard, nutlike (albeit tiny) fruits that aren't terribly fleshy at all, although developmentally they start out with a thin fleshy covering. One last trait of the common ancestor of the Cannabaceae is prophylar (or vegetative) buds. These are small bud-like protrusions that lie at the point where the leaf stem emerges from the main stem. But again, *Humulus* lacks these buds (or at least they aren't obvious). This doesn't mean the Cannabaceae ancestor didn't have the veinless teeth, fleshy drupe seeds, or prophylar buds. As both *Cannabis* and *Humulus* diverged from the common ancestor of Cannabaceae, these traits either could have been added (secondary veins and nutlike fruit in both *Cannabis* and *Humulus*) or lost (prophylar buds in *Humulus*).

To summarize: the common ancestor of Cannabaceae was vascularized with a circular bundled vascular system, had seeds, leaves that were broad with reticulate venation, flowers with parts in multiples of four or five, more than likely a hypanthium, leaves with serations, ascending veins that had teeth with no veins, drupe fruits, prophylar buds, and was dicotyledonous. It's starting to look like the modern cannabis plant we know and love.

4

Hops and Hemp

We have now circled back to *Humulus* and *Cannabis*—hops and hemp—with which we started chapter 3. We have a fairly distinctive-looking plant with the ancestor of Cannabaceae. But when we examine *Humulus* and *Cannabis,* we see further distinctions compared with other members of the family Cannabaceae. It is clear from genetic studies that these two are each other's closest relative, or what taxonomists call sisters or sister taxa. What did the ancestor of hops and marijuana look like? There are eight interesting morphological alterations that shaped the Cannabaceae ancestral morphology. These are shown in table 4.1.

More Big Events

Of these eight morphological changes, three are of particular interest for the *Cannabis/Humulus* ancestor: triporate pollen grains, imbricate flower aestivation, and a persistent perianth (fig. 4.1). A triporate pollen grain is one with three tiny pores on the surface of the grain. Imbricate flower aestivation occurs when the sepals or petals of a flower overlap or branch out separately. A persistent perianth is the part of the flower that is left when the rest of the flower dies and falls away.

Table 4.1. THE EIGHT MORPHOLOGICAL CHARACTERS AND THEIR STATES RELEVANT
TO *HUMULUS* AND *CANNABIS*

Morphology	States
Sex system	0 = monoecious; 1 = dioecious; 3 = monoecious or dioecious; 4 = polygamous
Leaf arrangement	0 = opposite; 1 = alternate; 2 = alternate and opposite
Pollen aperture	0 = triporate; 1 = diporate; 2 = pentaporate
Aestivation	0 = valvate; 1 = imbricate
Fruit type	0 = drupe; 1 = achene; 2 = samara
Seed coat	0 = with holes; 1 = without holes
Perianth at fruit	0 = deciduous; 1 = persistent
Stipule arrangement	0 = intrapetiolar; 1 = extrapetiolar; 2 = interpetiolar

Note: Characters in boldface are the three that discriminate *Humulus* from *Cannabis*.
Source: Yang et al. (2013).

The other five characters are not present in both of the *Humulus/Cannabis* sisters, but they are present in one or the other and hence do not help in defining what characters *both* have.

Our *Cannabis/Humulus* ancestor now has three new traits that we can append to it that distinguish it from all other relatives, indeed most all other plants. We can also use the formal name for cannabis plants based on the reconstruction of these ancestors. But before we give cannabis its entire formal name, we need to refer to a higher category called a domain. The domain in biology is simple because there are only three: Bacteria, Archaea and Eukaryota. This requires amending our mnemonic too; I like **D**evout **K**ing **P**hilip **C**ame **O**ver **F**rom **G**reat **S**pain, as historically the King Philip in the mnemonic could be King Philip II of Spain who married Queen Mary I of England in 1554.

We know that cannabis plants are eukaryotes (Domain: Eukaryota), and they are organisms located in the Kingdom Plantae. There are several phyla of plants, but which one is used depends on which of the ancestral traits previously discussed are most important to the person doing the classification.

Moving on from a single-celled lifestyle is a major evolutionary step, and this would peel off the green algae or the Chlorophyta, as they are technically

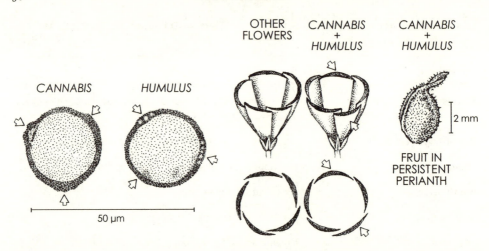

Figure 4.1. Three anatomical characters that are uniquely found in both *Humulus* and *Cannabis*. The left shows the triporate pollen grains of both genera. Arrows point to the three pores on the grains. The middle diagram shows the imbricate flower aestivation character, compared with the spiral aestivation character. The arrows point to the two petals in the imbricate form that throw off the spiral found in other members of the Cannabaceae. The right diagram shows the fruit encased in a persistent perianth. The perianth is the nonreproductive part of the plant that forms an envelope around the sex organs. It consists of the flower's corolla and calyx, which usually fall off the plant during maturation.

known. The next critical category is being vascularized or not; this distinction forces us to call mosses something important at the phylum level, and indeed these are placed in the phylum Bryophyta. Splits within the vascularized plants are the next big divergence. Ferns took off from the common ancestor that evolved vascularization. Ferns are then placed into the phylum Pteridophyta. Seeds were the next significant variable, so we would want to make sure the two large groups splitting here (angiosperms and gymnosperms) get decisive names at the phylum level. Therefore we name the gymnosperm lineage either Gymnospermae or Coniferophyta, and the angiosperm lineage Angiospermae or Magnoliophyta. We can combine our steps through the closer ancestors of cannabis to derive the following classification for cannabis, making King Philip very happy. At this point, there are three potential cannabis species: *sativa, indica,* and *ruderalis.* As we will see, this might not be the best way to categorize cannabis plants.

Domain: Eukaryota
Kingdom: Plantae

Phylum: Angiospermae
Class: Dicotyledonae
Order: Rosales
Family: Cannabaceae
Genus: *Cannabis*
Species: *sativa* (*indica, ruderalis*)

There are more complex classifications for cannabis plants, which move away from **D**evout **K**ing **P**hilip **C**ame **O**ver **F**rom **G**reat **S**pain. The classification shown below omits phylum and provides more detail in the transition from Plantae to Angiospermae by adding the two higher taxa, Tracheophyta and Pteropsida. Tracheophytes represent the evolution of the vascular system, and Pteropsida represent the evolution of seeds.

Domain: Eukaryota
Kingdom: Plantae
Division: Tracheophyta
Subdivision: Pteropsida
Class: Angiospermae
Subclass: Dicotyledonae
Superorder: Dilleniidae
Order: Urticales
Family: Cannabaceae
Genus: *Cannabis*
Species: *sativa* (*indica, ruderalis*)

How much information you want to impart to a classification dictates whether you prefer the King Philip version or the more detailed version. We will discuss many of the traits mentioned here, especially the attributes of the flowers of *Cannabis,* when we get up close and personal with the plant in chapter 5. For now, the anatomy of plants in general and *Cannabis* specifically have served us well in tracing its long journey to becoming what it is.

Names before Names

Names are important in situations where people live together. Clarity, speed, and imagination are probably the main reasons we give things names. Imagine that you and your Neolithic buddy are hunting in a forest. You come across a huge bear. It is much quicker to yell "Bear!" to warn your comrade than to say, "It's a big hairy animal with claws and a nasty disposition toward humans!" Or if you are about to chomp down on a delicious-looking red berry, by the time your buddy says, "It's that plant with red berries that makes your stomach turn to fire and kills you!" you may already be dead.

I have gotten a bit formal with names, but now I need to slip back into the vernacular. This slide back is necessary because the formal name of cannabis is somewhat controversial and confusing. Stepping back to the vernacular names enables us to probe the heart of a taxonomic controversy, but it also unveils more confusion.

There are about 6,500 languages in use in the world today. There are another 600 or so that linguists know have gone extinct in historical time. But that isn't the limit of the number of languages that our species has created in our 200,000-year existence (our common ancestor with Neanderthals is a little earlier, but 200,000 years ago is a good estimate for when modern humans arose based on the fossil evidence). Linguists place this upper estimate at around 31,000 human languages that have existed through time (but not all simultaneously). Not all of these 31,000 languages would have had a word for cannabis, because cannabis, before it recently gained its worldwide distribution, was what is called a restricted endemic plant. In areas where cannabis didn't exist, there was obviously no need for a word to describe it.

But cannabis was inserted into so many cultures that keeping track of the names is daunting. Cannabis plants have hundreds of names worldwide. We have already discussed the origin of the modern English word (cannabis) in chapter 2. The internet provides many of the names of cannabis and to list them all here would be a waste of ink. But some of the names remain in dominant use, such as "hemp" and "ganja." "Can-sa" and "hashish" are the next most common traditional names used, shown in the word cloud in figure 4.2. Slang terms such as "kush," "instaga," and "dody" are popular on the internet.

To grasp the evolution of words for cannabis, we need to understand how, where, and when cannabis came to be spread across the planet. The history of the plant's more modern movement is well understood, but we need to distinguish among its uses as medicine, as fiber, and as recreation, because its distribution was governed by these three uses. As the map in figure 4.3 illustrates, there have been four critical ages involved in the spread of cannabis. The first age (pre-Christ), when the plant itself was initially endemic, included what is now South-

Figure 4.2. Word clouds showing ancient culture names for cannabis (left) and colloquial names for marijuana (right).

east Asia, India, and China. ("Endemic" means cannabis was found there and nowhere else at the time. For instance, humans were initially endemic to Africa.)

The use of cannabis as a medicinal preceded the birth of Christ by centuries but was localized to southern and south central Asia. The Persians, Assyrians, and Scythians all are known to have used cannabis as a recreational or medicinal plant, and some of the older words for cannabis come from languages used in those regions. But even before human use, its endemism was only in restricted areas in Asia where it naturally occurred.

After a successful stint as a medicinal and a fiber crop plant for a couple of millennia, cannabis was introduced to the Arabian Peninsula in the tenth century and then to Africa and South America in the fifteenth and sixteenth centuries. Its distribution as a medicinal expanded to North America and Europe in the eighteenth century. It was also moved to Australia in the late 1700s when Europeans migrated to that continent, but its importation there was intended as a fiber crop. I want to add one more age that corresponds to the modern dispersal of cannabis as a major recreational drug, such that the entire globe (including all the light areas of the earth in fig. 4.3) now can be thought of as having marijuana. In little under 5,000 years, the plant went from being a restricted endemic in South Asia to a global weed in some cases, a medicinal in others, and a fiber crop in still others.

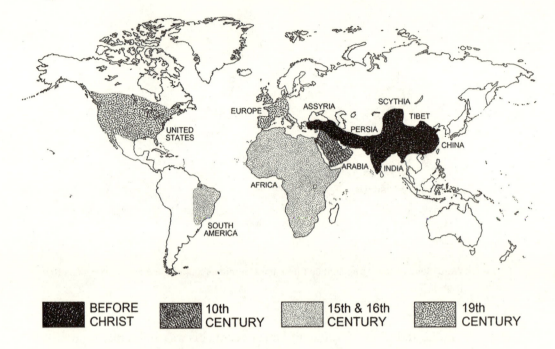

Figure 4.3. Four critical ages involved in the spread of cannabis across the globe.

Confusion Abounds, Simplicity Wins

It might be obvious from our discursion into the evolution and taxonomy of *Cannabis* that there is a certain degree of arbitrariness or subjectivity to the whole business. But what about the species in the genus *Cannabis*? Shouldn't it be easy to recognize different species? Species should be entities where reproduction has ceased between the entities, preventing them from genetic exchange. Species don't need to live in different places for such isolation (a situation biologists call allopatry). They can also live in the same place (called sympatry), becoming isolated enough genetically through other mechanisms to be considered good species. The major point is that reproductive isolation in some way exists to derail genetic mixing. This barrier to sex is followed by anatomical, chemical, or genetic differentiation, so that the two new entities have traits that can be used to diagnose their separateness.

But this neat definition of species is easily rendered useless. My thesis advisor Alan Templeton juxtaposed the phrases "too little sex" and "too much sex" in his lectures to describe the problems of using reproductive isolation as a speciation producing process. Too little sex is what we see in bacteria, archaea, and other clonal organisms. These organisms don't have sex (see chapter 3), and the reproductive isolation criterion can't be used to define species for these organisms, thus excluding perhaps 99 percent of all organisms on Earth as immune to that "isolation" definition. Too much sex refers to organisms that persist in mating even after isolation has occurred. In these cases, the species may have perfectly good diagnostic traits but can still interbreed with each other. Plants are notorious for having "too much sex" in the context of species definitions. In fact, many plants are—simply put—sex machines. There are cases in plants where species have mated across two different genera and produced viable and fertile offspring. Regardless, species remain an important part of giving an organism its formalized name. It has to be a species to warrant a species epithet. And species confusion is intricately related to how many formal names we give marijuana plants.

The Cannabis Name Game

There are about 300,000 species of plants that have been named, with another 75,000 still expected to be discovered and formally named. Carl von Linné (Carolus Linnaeus) was the first to give standardized binomial names to plants in his *Systema Naturae,* first published in 1735. This volume went through several editions and ended with the twelfth edition, published from 1766 to 1768. The tenth edition is perhaps the most famous of the dozen that Linnaeus produced, and it marks the start of modern taxonomy. All priority (which became an important term for naming things) traces back to this edition, which was actually titled *Systema naturæ per regna tria naturæ, secundum classes, ordines, genera, species, cum characteribus, differentiis, synonymis, locis* (System of nature through the three kingdoms of nature, according to classes, orders, genera, species, with characters, differences, synonyms, places). Linnaeus was basically working with a Came Over From Great Spain system then. Of the currently named 300,000 plants, Linnaeus named about 5,900 of them, and named 12,000 plants *and* animals during his half-century career. On average he named over 100 new species of plants a year (one every three days) during his career. *Cannabis sativa* was one of them.

Linnaeus was a stickler for precision and order. His binomial system (actually not entirely his, as natural philosophers before him had thought about binomial names) has lasted almost 300 years, and a grand majority of the names he published are still valid and in use today. *Cannabis sativa* is one of them and first appeared in Linnaeus's *Species Plantarum* (1753), which marks the birth of plant taxonomy. Another famous scientist, Jean-Baptiste Lamarck, entered the picture in 1785 by naming a second *Cannabis* species: *Cannabis indica*. About a century and a half later, in 1924, Russian botanist D. E. (Dmitrij Erastovich) Janischewsky described what he believed was a third species, *Cannabis ruderalis*. So in this scheme there is *Cannabis sativa* L., *Cannabis indica* Lam., and *Cannabis ruderalis* Janisch. The L., Lam., and Janisch. parts of the names refer to the abbreviated names of Linnaeus, Lamarck, and Janischewsky, respectively, their legitimate describers. The three plants do look different, as everyone knows who has grown or seen them (fig. 4.4).

These three entities have been referred to as species by their full scientific names or by three common or vernacular names: Sativa, Indica, and Ruderalis. Note that the scientific names are italicized, with the species name in lowercase, but the vernacular names are capitalized, not italicized, and omit the genus name. Sounds simple, but the official accepted names of these plants have been a jumble.

While some names below the species level can gain acceptance in scientific circles, the species epithet is the one that is sought when doing taxonomy. Cannabis has had a strange history due to its domestication and because most people cultivating the plant were not taxonomists. As is the case with many domesticated species, the vernacular or folk names stuck with the plant: "Sativa," "Indica," and "Hybrid" ("Ruderalis" has often been an afterthought). When names such as "Sativa" and "Indica" are used in common-language descriptions of *Cannabis* and in marketing cannabis products, these tend to stick in people's minds and confuse the issue with the potential validity of scientific species epithets such as *sativa* and *indica*. Sorting out the names has been a decades-long trek.

William Stearn and Richard Schultes and colleagues all contributed taxonomic treatments of the genus *Cannabis* to Vera Rubin's *Cannabis and Culture* book. Both chapters were less than impressed with the state of taxonomy of the genus in 1975, with one set of authors calling the genus "an example of taxonomic neglect." Table 4.2, partially compiled from Schultes and colleagues' work, shows the many names and their status in 1975.

Some of the names in the species list could be eliminated by being more careful about the specimens already collected. In these cases, a name such as *Cannabis macrosperma* is considered synonymized with a previously validated name.

Figure 4.4. Drawings of the entire plant and leaves of Sativa, Indica, and Ruderalis.

There are many cases of this in *Cannabis* nomenclature: *macrosperma, generalis, gigantea, intersita,* and *kafiristanica* all are considered synonyms of *Cannabis sativa.* These names are retained as synonyms and are not used to describe what is known as *sativa.* They have been shed from the formal taxonomy because of the rule of priority, which states that if any two names for the same thing are in the literature, the one that appeared first is the valid name.

Many of the species' names mentioned by Schultes and colleagues can be

Table 4.2. THE MANY NAMES GIVEN TO *CANNABIS* SPECIMENS FROM LINNAEUS (1753) TO SANDS (2002)

Species name	Variant name	Year	Author	Status
sativa		1753	Linnaeus	1975?
lupulus		1772	Scopoli	1975? (*Humulus*)
indica		1783	Lamarck	1975?
foetens		1792	Gilibert	nomen illegitimum
eratica		1796	Sievers	nomen nudum
macrosperma		1812	Stokes	1975? (synonym: *sativa*)
himalyana		1898	Zinger	nomen nudum
generalis		1905	Krause	1975? (synonym: *sativa*)
americana		1908	Haughton	nomen nudum
madagascar		1909	Pearson	nomen nudum
ruderalis		1924	Janischewsky	1975?
gigantea		1927	Crevost	1975? (synonym: *sativa*)
mexicana		1931	Stanley	nomen nudum
africana		1935	Wehmer	nomen nudum
pedemonata		1936	Camp	nomen nudum
intersita		1960	Sojak	1975? (synonym: *sativa*)
kafiristanica		1981	Chrtek	2020? (synonym: sativa)
afghanica		1987	Clarke	nomen invalidum
	vulgaris	1869	de Candolle	2020?
	montana	1869	Fristedt	nomen nudum
	hymalaiensis	1873	Cazzuola	homonym
	himalayensis	1876	Cazzuola	nomen nudum
	asperrima	1879	Regel	2020?
	crispata	1886	Hasskarl	2020?

Continued on next page

Table 4.2. *Continued*

Species name	Variant name	Year	Author	Status
	afghanica	1929	Vavilov	nomen nudum
	kafiristanica	1929	Vavilov	2020?
	culta*	1940	Serebriakova	2020?*
	asiatica*	1940	Serebriakova	2020?*
	narcotica*	1940	Serebriakova	2020?*
	turkistanica	1987	Clarke	nomen invalidum
	spontanea	1987	Clarke	nomen invalidum
	afghan	2002	Sands	nomen nudum

Sources: Schultes et al. (1975) and McPartland (2020).
* = below variant.
1975? = status cloudy in Schultes et al. (1975).
2002? = taxon sunk in McPartland (2020).

disregarded because of *nomen nudum, nomen invalidum,* or *nomen illegitimum* issues. *Nomen nudum* ("naked name") simply means insufficient data were offered for the name. *Nomen invalidum* ("invalid name") is a name that was in use but was not supported by a valid publication. *Nomen illegitimum* ("illegitimate name") is a validly published but illegitimate name, usually because its naming did not follow the rules of the governing body for plant taxonomy in one way or another. *Nomen invalidum* cases can happen when a taxonomist comes up with a name and advertises it to his colleagues but does not validly publish it, usually because of a desire to wait until a complete understanding of a group is reached. *Nomen nudum* and *nomen illegitimum* are labels of taxonomic laziness. *Nomen nudum* is lazy because one simply didn't write the paper to describe the species; *nomen illegitimum* is lazy because the author of the description didn't do the scholarly work to determine whether the specimen described already had a prior name. By the way, an illegitimate name can never be used again in the genus where it was first used. So, for instance, we will never again see the *Cannabis foetens* epithet for a marijuana species. Even if more species in the genus *Cannabis* are discovered that name cannot be used for a new *Cannabis* species.

Lump, Split, Twist

Earnest Small is an agricultural specialist in Canada who first entered the cannabis research area in 1970, when he worked for the Canadian government. In 1976
he and one of the gurus of plant systematics, Arthur Cronquist, attempted to clarify aspects of *Cannabis* nomenclature. Another cannabis researcher who jumped
into the name game is John McPartland, who has made a career out of concern
for psychoactive plants. A family doctor in Vermont and a professor at the University of Vermont, he has studied marijuana, its history, its neurological and medicinal effects, and especially its name for the past forty years. Both Small and
McPartland recognized that an officially accepted name for the plant is extremely
important not only for science but for legal and medical aspects. Together in 2020,
they finally published the most comprehensive treatment of cannabis taxonomy.

McPartland and Small play quite rigidly by the rules of taxonomy established by the International Code of Nomenclature for algae, fungi, and plants
(ICN or ICNafp). There are eight rules that cover almost every imaginable aspect
and loophole in officially naming something. Particularly important to Small and
McPartland was a rule governing type specimens. They believed that if they could
untangle the sordid history of where and how marijuana specimens have been
named, they could lend some sense to the names given to marijuana plants. This
wasn't a trivial matter, because plants in the genus *Cannabis* have had at any given
time perhaps two dozen binomials leveled at what might actually be only a single
species.

Stearn, Schultes, and colleagues had contributed their understanding of the
taxonomy of the genus in 1975. A year later, Small and Cronquist developed their
own taxonomy that simplified much of the knottiness of *Cannabis* taxonomy.
Cronquist, one of the giants of plant taxonomy, developed a system of plant classification called the Cronquist system. He also was the coauthor of the "green
bible" (*Manual of Vascular Plants of the Northeastern United States and Adjacent
Canada*), a comprehensive identification guidebook. Cronquist was known for
his significant accomplishments as a plant taxonomist but could not avoid being
called a "lumper." There are three kinds of taxonomists: lumpers, splitters, and
those who do neither or both. Splitters are taxonomists who choose to name lots
of species to account for the variations they see in groups; they like large numbers of species. Lumpers prefer to designate fewer species, with larger population membership in each. True to their reputations, Small and Cronquist simply
called *Cannabis* a single species (*Cannabis sativa*) with two subspecies—*sativa*

and *indica*. This worked for Cronquist because of his lumpy nature (both physically and taxonomically) and because, as Small and Cronquist stated in their 1976 treatment, "there appear to be no barriers to successful hybridization within the genus," which agrees with the species definition given above.

So now there was a second scheme: instead of three species, there was one species. It is the job of taxonomists to resolve such conundrums, which is what Small and McPartland set out to do. They took on the nomenclature of *Cannabis* with an exhaustive search of all the binomial names used by researchers since Linnaeus started it all. They scoured the literature and the herbaria of the world for scientific papers and specimens relevant to the taxonomy of *Cannabis*. Their exhaustive search of the genus found over 1,100 specimens in fifteen different herbaria across the globe. They discovered more than two dozen additional names used in *Cannabis* taxonomy (see table 4.2), many of them levied at taxonomic levels lower than species (i.e., subspecies and variants). These new names made the mess a bit messier, but they were important to scrutinize because they could have been authentic species.

One thing was for sure: the name *Cannabis sativa* L. was going to be kept in the scheme—you don't overturn nearly 300 years of a stable name, especially when the giant of taxonomy himself had levied it. But if there was only one species, then the species names *indica* and *ruderalis* were in deep trouble. And if those names were somehow sunk into *sativa,* they could never be used again to name another species in the genus if one were ever found.

Of all of the names listed, one could be eliminated right away: *Cannabis lupulus*. Note that the species name in the epithet is identical to the species name of the hops plant, *Humulus lupulus*. The author of the work, Giovanni Scopoli, used this name because he wanted to draw attention to the close similarity of marijuana and hops, but he didn't do the cannabis name game any favors by doing so. Scopoli's gaffe was easily sorted out by taxonomists, because no one accepted the idea that hops and marijuana were in the same genus, even though there is now complete acceptance of the sister relationship of hops and marijuana. Small and McPartland could eliminate a few more of the other names quite easily by using the rules of the ICNafp. Most of the names were *nomen nudum,* meaning that no publication or scholarly justification existed for the name. Table 4.2 shows just three names—*Cannabis sativa, Cannabis indica,* and *Cannabis ruderalis* (all in bold)—that were not eliminated by these processes, and implying that these are indeed the three currently accepted and valid species in the genus *Cannabis* today. Game over. Or was it?

McPartland and Small suggested that one of the reasons we are left with these three epithets is because of the vernacular or "folk" use of the names Ruderalis, Sativa, and Indica. Because of this confusion, Small and McPartland justified splitting at the level of subspecies and not at the level of species. They doubted that the plants called Ruderalis or *Cannabis ruderalis* are a genuine species but rather are members of one species, *Cannabis sativa*. They suggested that the vernacularly named Indica plants are actually a variety of subspecies *indica* called *afghanica*. "In summary," they stated, "reconciling the vernacular and formal nomenclatures: 'Sativa' is really *indica*, 'Indica' is actually *afghanica,* and 'Ruderalis' is usually *sativa*. All three are varieties of one species, *C. sativa* L." This most recent taxonomic analysis of cannabis from Small and McPartland is reasonable given the species definition described above and the arguments involving DNA barcoding (see chapter 6). McPartland's way of dealing with this taxonomic problem is to lump and then split—in essence a recapitulation of Small and Cronquist but with some twists. He points out that the taxonomy from 1976 has never been replaced. It has simply been expanded.

C. sativa subsp. sativa var. sativa (low THC, with domestication traits)
C. sativa subsp. sativa var. spontanea (low THC, wild-type traits)
C. sativa subsp. indica var. indica (high THC, domestication traits)
C. sativa subsp. indica var. kafiristanica (high THC, wild-type traits)

This scheme leaves us with only one species (*C. sativa*) and two subspecies (*sativa* and *indica*). It is automatic when naming two or more subspecies within a single species to name one of them with the same species name as the subspecies name. That is why one of the subspecies names is *sativa,* giving us *Cannabis sativa sativa*. Likewise, when naming a variant, the species/subspecies epithet is used for one of the variants. Hence the name *Cannabis sativa sativa sativa* is valid for the first entity in the list above. Within the two subspecies posited by Small and Cronquist there are at least two varieties. Small and McPartland go further with the varieties within the subspecies *indica*. They designate four varieties within the subspecies *indica,* retaining the variety *indica* (*Cannabis indica indica indica*) and splitting Small and Cronquist's variety called *kafiristanica* into varieties *afghanica, asperrima* and *himalayensis*.

This scheme seems to go a long way to clearing up the confusion, but it may eventually fall because of the amount of hybridization and the immense number of cultivars that might need to be brought into the taxonomy. I myself am a lumper, and I prefer the single species name *Cannabis sativa* to describe all marijuana plants regardless of THC content or culturing process. I do not think subspecies fit within a diagnostic concept, and with taxonomy, species are justified via diagnostics. With species descriptions taxonomists will list all of the traits that can diagnose the entity. "Diagnose" simply means that a given trait is unique to the species being described. In essence, the diagnostics test the hypothesis that a group is a species. Subspecies do not have diagnostics or a particularly cogent way to view a subspecies as a hypothesis. In my view, a subspecies is essentially a flagged group of specimens that require further work to determine whether they are species. In this sense they are not particularly useful and are sometimes confusing. Another way to deal with this "splitting" in nomenclature is to use the term "cultivar." This is a common way that agriculturally important entities get recognized, and with cannabis there are thousands of these cultivars.

The subspecies epithet is only useful in guiding future work and not in resolving the true nature of cannabis nomenclature. A similar problem occurs with the vernacular name "hemp." This name has already been used in this book to describe strains of cannabis that are used in fiber technology. Hemp is also a name that is used in legal circles to describe a cannabis strain with less than 0.3 percent THC content. But that 0.3 percent boundary is rather arbitrary, and hence the name "hemp" is not only vernacular but also imprecise.

What is the solution? I believe we should settle on the existence of a single species for all of the current cannabis strains: *Cannabis sativa.* Once this taxonomic level is settled, then it remains for taxonomists and cannabis workers to get down to the business of systematizing the strains of cannabis that exist.

Because most of cannabis variants are the result of cultivation, their naming comes under a different code than the ICNafp code. And because naming strains in the marijuana business is so important, consideration of how to name these strains has gotten some attention. The official cultivar code is called the International Code of Nomenclature for Cultivated Plants (ICNCP). Very few cannabis strains, however, have met the ICNCP's nomenclature requirements, and these rules will need to be applied to cannabis strains in the future. While there may be a bit of a calm with respect to cannabis taxonomy as implied by Small and McPartland (except for what might be called excessive splitting after the subspecies level), it is doubtful that a "taxonomic storm" can be avoided given the thousands of strains that might need to be systematized by the codes of nomenclature. Stay tuned.

5

A Complicated Sex Life

How the cannabis plant develops from a fertilized ovule to a full-fledged reproducing organism is a fascinating biological story. How populations of this plant form part of a larger ecosystem and how cannabis, as a lineage, has changed through evolutionary time are also big parts of its life history. The most logical place to start is the ovule (which makes the egg) and the pollen grain (which produces the sperm), and then to work our way up to the mature plant with its leaves, stems, and flowers. Eventually its seeds and pollen start the cycle of development all over again.

Plant Porn

Flowering plants have diverse sex lives (fig. 5.1). To start, for flowering plants—that is, plants that make flowers, like cannabis—all sex occurs within that flower. In fact, the flower exists to facilitate effective reproduction with colors, scents, and nectar all evolved to attract pollinators to the flower and effect pollination, or the placing of the pollen in close proximity to the ovule so that sperm and egg can eventually meet to result in fertilization. The pollen is produced in structures

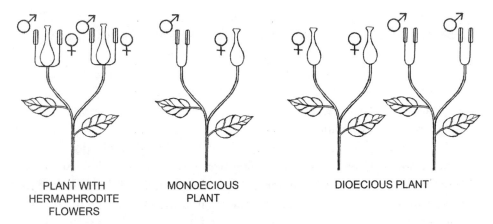

PLANT WITH   MONOECIOUS   DIOECIOUS PLANT
HERMAPHRODITE   PLANT
FLOWERS

Figure 5.1. Drawings of typical monoecious and dioecious flowering plants. There are many ways a plant can have its genitals structured. Very common among flowering plants are hermaphroditic flowers (left), where both the stamen and pistils develop within the same flower. In this case, the flower has both male (stamen) and female (pistil) reproductive structures. Other plants produce flowers that have either stamens (male flowers) or pistils (female flowers), but not both. These are called imperfect or unisexual flowers. For monoecious plants (middle), male and female flowers appear on the same plant. For dioecious plants (right), a male plant produces only staminate flowers, while a separate female plant makes only pistillate flowers. Adapted from Nefronus (Wikipedia drawing).

called stamens (see fig. 5.1, male symbol), while the ovules are produced in structures called ovaries within the pistil (see fig. 5.1, female symbol).

One aspect of the diversity of plant sex comes from that fact that some flowers make both pollen and ovules, while others make only one or the other but not both. As with humans and other animals, plants have two sexes that contribute to the reproduction of the species. The two sexes make different gametes: the sperm (made by males) and the egg (made by females). In plants, these two gametes can be produced by the same individual as part of a single flower. The majority of the over 350,000 named flowering plants are hermaphroditic, meaning the flowers produced have both stamen and pistils (see fig. 5.1, left panel). However, some plants are what are called monoecious ("one house"), where flowers are either staminate (male) or pistillate (female), and a single individual produces both male and female flowers (see fig. 5.1, middle panel). This state is more than likely the primitive state for flowering plants, meaning that the common ancestor of all angiosperms was monoecious.

Cannabis uses this way of organizing its sex organs. But in a supersexed evolutionary move, cannabis also features a different type of male/female sex system called dioecious ("two houses"), whereby some plants make only female (pistillate) flowers and others make only male (staminate) flowers (see fig. 5.1, right panel). About 7 percent of all flowering plants practice this kind of sex organization; compare this to 95 percent of all animals, which have a similar organization of the sexes called gonochorism.

To dig a little deeper into the sex life of cannabis, I should point out its male plant genitalia, the stamen (the pollen-bearing organ), is made of two major structures—the anther and the filament. The anther carries the pollen, and the filament simply connects the anther to the stem of the flower. The pistil of the female genitals is a bit more complicated, with four major subparts: the stigma, style, ovary, and ovules.

I have oversimplified the flower sex morphology and breeding systems of plants, because the names of the different breeding systems are daunting tongue twisters (see table 5.1) and because for cannabis we only need to know that most individuals are either strictly male or strictly female, although hermaphroditic flowers do arise spontaneously in some marijuana breeding populations.

Overall, plants are much more "pansexual" than they are strictly dioecious or monoecious. A fascinating aspect of these breeding systems is how they evolved. Evolutionary botanist Susanne Renner points out that because pollen and stamens appear in individuals for a large number of flowering plant lineages, bisexualism can be considered the primitive state for flowering plants. She also suggests that there were anywhere from 900 to 5,000 independent derivations of the different kinds of dioecy that exist. This means that there was a lot of evolution from bisexuality to dioecious breeding systems in plants.

If most cannabis plants today are dioecious, was the ancestor of all cannabis dioecious? One way to decipher this is to examine the genus *Cannabis*'s closest relative *Humulus* (hops). These plants are also mostly dioecious, with male and female flowers strictly on different individuals. Bisexual flowers do occur at a very low frequency. The most parsimonious interpretation, then, is that the breeding system of the hops/marijuana common ancestor may have been a dioecious plant with separate sexes. Application of different growth regulator chemicals to cannabis plants can alter the strict sexuality of marijuana flowers, and this result implies that it is relatively easy to transition from a population of plants with two sexes to a population that includes bisexual plants. Physical and chemical stress can also induce the appearance of male-like flowers on female cannabis plants.

Table 5.1. THE SEVERAL KINDS OF PLANT SEX

Name	Abundance	Description
Monocliny	85 percent	Perfect bisexual plants; flowers are truly bisexual
Distyly	Rare	Bisexual size of the male parts and female parts vary
Gynomonoecy	Rare	Female and hermaphroditic flowers on same individual
Gynodioecy	Rare	Female and hermaphroditic plants in same population
Monoecy	< 7 percent	Male flowers and female flowers on same plant
Dioecy	< 6 percent	Separate male and female individuals
Andromonoecy	Rare	Male and hermaphroditic flowers on the same plant
Androdioecy	Rare	Male and hermaphroditic plants in same population

Stressors such as disrupted photoperiods and low temperature may increase the formation of male flowers on female plants. That these environmental and chemical stressors can spontaneously induce hermaphroditism offers another bit of evidence for the primitive dioecious state of *Cannabis* and *Humulus*. The only factor that is critical for the survival of a plant species such as cannabis is that there are some male flowers and some female flowers in the population that will produce pollen and ova. Seeds are produced by ova coming together with pollen. It shouldn't matter whether these flowers are on the same plant (monoecious) or on different plants (dioecious). But this would be a naive assumption about the sex lives of plants, for some plants can do "virgin birth" or parthenogenesis, when seeds are produced without the act of fertilization. This brings us to a curious little debate from the 1800s as to whether cannabis can reproduce parthenogenetically.

Apparently Lazzarro Spallanzani of anti–spontaneous generation fame also worked on cannabis, or what he called hemp. An Italian abbot, he was considered a careful scientist and was responsible for debunking spontaneous generation (the hypothesis that living organisms could develop from nonliving matter) in a series of experiments. He tried similar experiments with plants, and it appeared as if cannabis could make seeds without fertilization of the ovule by pollen-producing plants. In the 1800s French botanist Charles Naudin repeated Spallanzani's experiments with cannabis and found the same results. Naudin also came within a hair's breadth of discovering Mendel's principles of inheritance several decades before Mendel. If he had only thought of doing crosses the way Mendel did, then the principles of segregation and random assortment might be called Naudin's laws instead of Mendel's laws. In 1861 Hermann Karsten repeated the observations of Naudin in a research paper and offered a much simpler explanation for Naudin's experimental results: "The researches of Naudin were instituted on polygamous plants—a circumstance which naturally suggests to the mind that a concealed male flower, or an anther produced in the interior of a female flower, may have led the observer into error."

Karsten was tough. In the 1861 paper he debunked several claims of virgin birth in plants with relish. Here is his statement about researchers of his time who claimed parthenogenesis occurred in plants: "Still, the propensity to credit what is marvellous, and to excite an interest by taking up the defence of bold hypotheses at variance with hitherto acknowledged laws, did not allow the results arrived at by the united assiduous labours of so many naturalists to go unchallenged." In other words, if you don't have the hard evidence, don't bother us serious scientists with your unsupported hyperbole.

Although Karsten was a harsh critic of parthenogenesis in plants, he did acknowledge its occurrence in algae and other lower plants. Today parthenogenesis has been shown to occur in approximately 400 plant species, which would probably shock Karsten. But he was right on one of these occasions: to date, virgin birth in cannabis has not been validated despite the excitement over early experiments that suggested its possibility. Today's researchers are similarly excited about the possibility that genes responsible for parthenogenesis might be useful targets for genetic engineering in cannabis. Imagine a plant like cannabis that can reproduce parthenogenetically. That would be a cannabis breeder's dream—no male flowers needed and only buds (female flowers) for generation after generation.

The pansexuality of cannabis is important in its commercial growth. Perhaps not so much for those interested in growing hemp, but certainly for growers interested in the recreational and medicinal aspects of marijuana. When a female plant's ovules are fertilized, its flowers set seed, which is an undesirable event for recreational and medicinal cannabis. Growers will hunt for males (staminate flowers) in their crops and remove them to prevent fertilization and subsequent seed production by the female plants in the crop. Certainly, any hermaphroditic flowers are problematic too—especially if some of the male genitalia are sneaky, such as those that fooled Spallanzani and Naudin centuries ago.

Warehouses—Beyond Sex

So the sex life of cannabis is complex. But to say that the development of the cannabis plant itself is complex is an understatement. To simplify or codify its development, agronomist Vito Mediavilla and colleagues devised a digital system for designating the developmental stages of the cannabis plant (table 5.2). Basically, there are four major developmental stages:

- Stage 0000   Germination and emergence
- Stage 1000   Vegetative stage
- Stage 2000   Flowering and seed formation
- Stage 3000   Senescence

Each of the base stages (0, 1, 2, and 3) can be subdivided into further substages by the addition of numbers in the second, third, and fourth positions. So 1100 refers to stage 1, substage 1.

Most cannabis researchers and enthusiasts recognize these stages in some way, but what Mediavilla and colleagues managed to do was categorize the developmental stages using a digital coding system, enabling a more precise description of cannabis development. In this chapter we will look closely at stage 0 and stage 1, where cannabis's developmental program is kick-started.

The unfertilized ovule and the pollen grains are strange kinds of cells compared with the rest of the cells in the adult cannabis plant. Adult plant cells not involved in reproduction have two copies of every gene in their genome and are called diploid. The cells involved in reproduction—pollen and ovules—have a single copy of their genes and are called haploid. A seed, otherwise known as a fertilized ovule, is the product of a single pollen tube penetrating the ovule's outer membrane and releasing a sperm cell that fuses with the ovule's egg. The pollen's haploid genome (via the sperm) is released into the interior of the ovule, where the ovule's haploid genome (via the egg) fuses with the sperm to form the nucleus of a single fertilized diploid cell. This diploid cell is called a zygote, which will develop to become the embryo and eventually the seedling. In this sense, the overall nuclear genome of a diploid organism is formed by biparental inheritance (one-half of the new genome from the pollen/sperm and one-half of the genome from the ovule/egg). Here again, plants are notorious for violating this mode of diploid union by increasing the number of chromosomes or genomes that reside in their parents. These kinds of increases cause what is called polyploidy.

Table 5.2. MEDIAVILLA AND COLLEAGUES' DIGITALIZED DEVELOPMENT SCHEME FOR CANNABIS

Code	Description	Remarks
Stage 0	Germination and emergence	
0000	Dry seed	
0001	Radicle apparent	
0002	Emergence of hypocotyl	
0003	Cotyledons unfolded	
Stage 1	Vegetative stage	Refers to main stem; leaves are considered as unfolded when leaflets are at least 1 cm long
1002	1st leaf pair	1 leaflet
1004	2nd leaf pair	3 leaflets
1006	3rd leaf pair	5 leaflets
1008	4th leaf pair	7 leaflets
1010	5th leaf pair	
Stage 2	Flowering and seed formation	Refers to the main stem, including branches
2000	GV point	Change of phyllotaxis on the main stem from opposite to alternate
2001	Flower primordia	Sex nearly indistinguishable
Dioecious plant—male		
2100	Flower formation	First closed staminate flowers
2101	Beginning of flowering	First opened staminate flowers
2102	Flowering	50% opened staminate flowers
2103	End of flowering	95% of staminate flowers open or withered

Continued on next page

Table 5.2. *Continued*

Code	Description	Remarks
Dioecious plant—female		
2200	Flower formation	First pistillate flowers; bract with no styles
2201	Beginning of flowering	Styles of first female flowers
2202	Flowering	50% of bracts formed
2203	Beginning of seed maturity	First seeds hard
2204	Seed maturity	50% of seeds hard
2205	End of seed maturity	95% of seeds hard or shattered
Monoecious plant		
2300	Female flower formation	
2301	Beginning of female flowering	First pistillate flowers; perigonal bract with no pistils
2302	Female flowering	50% of bracts formed
2303	Male flower formation	First closed staminate flowers
2304	Male flowering	Most staminate flowers open
2305	Beginning of seed maturity	First seeds hard
2306	Seed maturity	50% of seeds hard
2307	End of seed maturity	95% of seeds hard or shattered
Stage 3	Senescence	
3001	Leaf desiccation	Leaves dry
3002	Stem desiccation	Leaves dropped
3003	Stem decomposition	Bast fibers free

Cannabis generally avoids polyploidy, and in the wild it is found only as a diploid. Most cultivars to date are also diploid. But there are advantages to polyploidization. In domestic plants such as wheat and barley, random ancient polyploidization events have been essential to the age-old cultivation and popularity of these plants. In modern plant breeding the process of polyploidization can overcome species boundaries and allow interspecific hybrids to form—that is, new plant species can be formed by the successful reproduction of two separate species that maintain the genomes of both parental plants. Such polyploid hybrids are often important in plant breeding because they produce novel plant phenotypes that neither parent is able to produce. Polyploidization is also important in generating seedless cultivars (such as bananas or grapes) and can increase the resistance or tolerance of a plant to drought or poor soils. It is no surprise that agronomic research on the effects of polyploidy in cannabis has recently begun. Using various chemicals known to generate polyploid cells, Jessica Parsons and colleagues have found ways to produce tetraploid (four genomes) cannabis. Their research shows that leaves of tetraploid plants are larger than those of diploids and yet only half as dense. Most important, CBD content increases by an average of 9 percent in the tetraploid plants, but there is no observed increase or decrease in THC content. Tetraploid cannabis also has significant alterations of terpene profiles, which are important characteristics of both recreational and medicinal cannabis. As such, polyploid cannabis could produce a rich source of variation for cultivated marijuana.

Both pollen and ovules have two small organelles called mitochondria and chloroplasts in their cytoplasm, as discussed in chapter 3. Chloroplasts are involved in photosynthesis, which produces food in the form of sugars for the plant, and mitochondria are involved in respiration, a key component of energy production. Chloroplasts and mitochondria contain small circular pieces of DNA that are referred to, respectively, as the plastid (chloroplast) and mitochondrial genomes. Pollen grains do not tend to have chloroplasts because they do not undergo photosynthesis and are quite small and short-lived. As a result, the sperm itself has a nuclear genome from the parent plant but often does not bring with it a mitochondrial or a plastid genome. On the contrary, the egg has both a mitochondrial and a plastid genome, since both organelles are present in the ovule. The mitochondria and chloroplasts are inherited in effectively uniparental or clonal manners—with only maternal inheritance of their organellar genomes.

In order for a single fertilized egg to develop to a full-fledged cannabis plant with billions of cells, it obviously needs to produce more cells. The unfertilized ovule, with its egg nucleus inside, acts as a warehouse for DNA, encoding the proteins responsible for growth but also for many proteins that the ovule has manufactured to ensure that the seedling survives. Unfertilized ovules cannot make more cells, because they don't have the right molecular signals until they are fer-

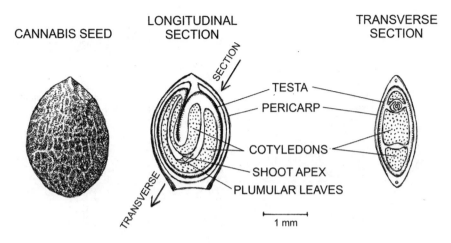

CANNABIS SEED

LONGITUDINAL SECTION

TRANSVERSE SECTION

TESTA
PERICARP
COTYLEDONS
SHOOT APEX
PLUMULAR LEAVES

1 mm

Figure 5.2. Drawing of a typical cannabis seed (left) with interior parts labeled on a longitudinal section in the center and a transverse section labeled on the right. Redrawn from Helga Mölleken and Roland R. Theimer, "Survey of Minor Fatty Acids in *Cannabis sativa* L. Fruits of Various Origins," *Journal of the International Hemp Association* 4, no. 1 (1997): 13–18.

tilized. The signals to start cellular division occur when the two haploid genomes of the sperm and egg interact with each other to form the diploid genome of the developing zygote. Once the signals are right, the fertilized ovule starts to divide and to produce new patches of cells and structures (fig. 5.2). This will eventually become the seed, with the embryo inside.

Once fertilization has occurred, a seed is formed, inside of which the embryo develops. It develops to a point and then stops and waits for germination. Germination is when the embryo inside the seed will burst out from the seed coat and establish itself as a seedling.

The seed is composed of a seed coat or pericarp, along with the developing embryo inside. The embryo includes rudimentary leaves called cotyledons (fig. 5.3); these first leaves do different things in different species, but always with the goal of nourishing the developing seedling. In some, they emerge from the germinating seed to be the first leaves to undergo photosynthesis. In others, they stay inside the pericarp and become filled with embryo-nourishing starches (as with walnuts and peanuts). In both cases, a rich complement of proteins made by the mother plant resides in the seed. Estimates of the number of different kinds of proteins in the unfertilized ovule of cannabis range from about 168 to 181, depending on the cultivar and on the environments where the plants are found.

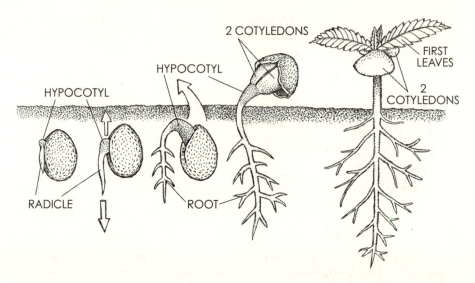

Figure 5.3. Drawings of the early events in germination of cannabis. The root radicle emerges first and grows downward, followed by the hypocotyl, which subtends the cotyledons. In cannabis the two cotyledons serve a function of both storage and photosynthesis. The first leaves are the earliest to resemble the mature/adult leaves of the cannabis plant.

Much of the research so far has been done on hemp cultivars (that is, strains with less than 0.3 percent THC concentration) because of legal constraints, so what has been observed from the current literature might be true only for hemp strains. But our basic knowledge of hempseed proteins can give recreational and medicinal cannabis breeders a good idea of the general makeup of seed proteins for future research.

The major proteins in cannabis seeds are what are called storage proteins, which support the embryo's development once it has germinated. There are four major kinds of seed storage proteins that were discovered nearly 100 years ago, and they are based on what kinds of solutions the isolated proteins can dissolve in: water-soluble proteins (albumins), dilute saline proteins (globulins), alcohol/water mixture proteins (prolamins), and diluted acid or alkali proteins (glutenins). These proteins are so important that the plant's genome has several copies of each, grouped into what are called multigene families.

The major seed storage protein in cannabis is a globulin that makes up about 80 percent of the seed's content. There are seven copies of this kind of gene called edestin that can be found in the hemp genome. The second most abundant category of seed storage proteins are albumins. These offer a good source of nutrition not only for the developing plant but also for any animal that eats the seeds, and they provide the developing embryo with its required nourishment until it can begin to photosynthesize and produce its own food.

The ungerminated cannabis seed carries no THC and doesn't start to produce THC until after germination, and at a very slow pace. By three weeks of age, though, the developing plant has produced enough THC to measure its concentration, so that its chemotype can be determined. Chemotypes will be discussed later in this book; suffice it to say here that at three weeks a typical marijuana plant will produce enough THC to tell if it will be above or below 0.3 percent THC concentration—the magic number for whether it is called hemp or recreational/medicinal marijuana.

oooo (Germination)

Everything that I have described so far in cannabis development is pre-stage oooo. The seed next needs to fall from the plant and to "set" or germinate (stage oooo). The germination of cannabis determines many outcomes for the plant in later life, so germination itself has been studied in some depth by hemp researchers. Much of this research indicates that success in germination depends on seed size and seed source. Different strains will respond to germination cues differently, and some of this difference is likely due to survival-based adaptations to different environmental stimuli. Pretreatment of ungerminated seeds with conditions such as changes in temperature or with chemicals such as plant growth regulators, disinfectants, and other exogenous chemicals can also significantly influence germination behavior.

Figure 5.3 shows the developmental steps in germination. Note that there are two directions that tissue will grow out from the seed: upward and downward. Downward go the roots, and upward goes the hypocotyledon. Some plants will develop the hypocotyledon while still underground; these plants are referred to as hypogeal. Other plants wait until the node that will eventually become the cotyledon is above ground. This way of growing is called epigeal; as figure 5.3 suggests, cannabis is epigeal.

The onset of germination is also recognized by a structure called a radicle poking out of and downward from the seed. The radicle will become the plant's root. This tiny projection responds to gravity and seeks out a place to anchor the developing seedling. Roots in cannabis (and in most plants) are of three kinds: tap roots, fibrous roots, and adventitious roots. A tap root is the main root that emerges directly from the radicle and grows into the ground as deep as it can. Fibrous roots grow off the tap root and provide additional support while reaching for soil-based nutrients. Finally, adventitious roots will emerge from the parts of the plant above ground, such as the stem. These roots reach down into the substrate for additional support and nourishment. They can be removed and planted apart from the original plant, providing growers with a way to clone cannabis plants.

Charles Darwin believed that the radicle was the "brain" of a plant. Although some recent research programs have been centered around so-called neural behavior in plants, the similarity to animals' nervous systems is just that—similarity. Plants do not have brains or nervous systems like ours that act as a single control center.

Although you might think that life is boring for a root, you would be wrong. There are millions of microbes living in the soil that interact with plant roots at any given time (fig. 5.4). These microbes include fungi, algae, single-celled eukaryotes called protists, and, of course, bacteria. They are all part of a separate ecological domain of the cannabis plant, and all the interactions have profound impacts on its growth. The interactions that are the most important for cannabis roots involve rhizobacteria, endomycorrhizal fungi, and symbiotic fungi such as *Trichoderma*.

One of the basic roles of a root is to absorb nutrients from the soil in which a plant is growing. Plant roots in general and cannabis roots specifically have entered into symbiotic relationships with microbes such as *Trichoderma* and endomycorrhizal fungi to help speed the uptake of nutrients and promote healthy growth and flowering. Plant researchers who have noticed the importance of organisms such as *Trichoderma* call them "plant growth–promoting microorganisms" (PGPM). PGPMs also offer a possible way to bypass the overuse of fertilizers that can cause disruptions in soil health.

Such symbiotic arrangements are not, however, one-way only; the fungi also gain benefits from their association with cannabis roots. The fungi confer an advantage to the cannabis plant by enhancing the plant's ability to absorb nitrogen and phosphorus. The fungi benefit because the cannabis plant will exude nutritious compounds such as sugars produced from photosynthesis. One of the more important benefits cannabis gains from its association with these symbionts

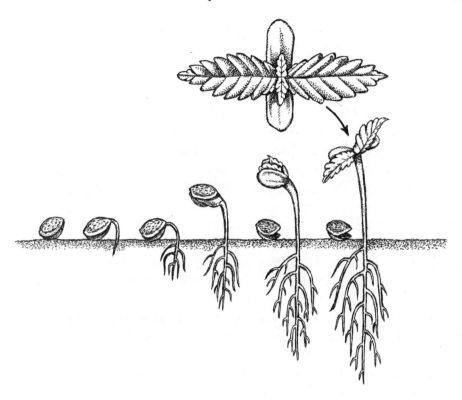

Figure 5.4. Drawings of the development of the root system of cannabis encompassing all of Stage 0 (0000, 0001, 0002, 0003). View from the top of stage 0003. Adapted from Artyom Kozhe-maykin, https://www.vecteezy.com/vector-art/1221027-stages-of-cannabis-seed-germination.

is protection from deadly fungi such as *Pythium* and *Fusarium*. Ioanna Kakabouki and colleagues have completed controlled experiments to examine the role of *Trichoderma* application to hemp plants. (Remember that hemp strains of cannabis may respond differently to experimental conditions than psychoactive/medicinal strains, but the results of the study are intriguing.) In hemp strains, flower production was upped by inoculating the roots of experimental plants with the fungus. A notable result of this study was that CBD content was also positively affected by the application of *Trichoderma*. More knowledge of the ecology and physiology of the microbial communities in and around cannabis roots is needed, and will undoubtedly reveal important aspects of microbial influence on the growth of the plant.

1000 (Vegetation)

From this stage on, the cannabis plant grows via cell proliferation of its main stem and by branching. This vegetative growth, known as stage 1 (see table 5.2), has five substages (1002, 1004, 1006, 1008, and 1010) defined by the sequence of leaf appearance on offshoots from the main stem. For instance, stage 1002 is defined by the emergence of the first two paired leaf offshoots, and stage 1010 is defined by the appearance of the fifth offshoots. The main stem, called the shoot apical meristem (SAM), contains the stem cells that can differentiate into different plant organs that appear above the ground. This shoot, the plant's critical backbone, gives rise to inflorescence meristems that can have two functions: vegetative meristems or floral meristems. Vegetative meristems produce the stems and leaves of the plant's branches. When conditions are right, the plant will enter the reproductive stage, and an inflorescence meristem will convert to a floral meristem, which ends with a flower. The branching (or ramification) of a branch or inflorescence meristem follows preprogrammed patterns.

During vegetative growth (fig. 5.5), the stem grows thicker and extends to be taller. It produces developmentally active regions called nodes from which branches extend. At first the nodes develop in pairs along the main stalk, but as the plant begins to mature the branches will start to alternate. Secondary nodes also develop on the nascent branches themselves, which increases the number of secondary branches and leaves. This increase facilitates greater photosynthesis by the leaves and provides more nourishment for the plant. As the plant grows, the nodes along the stem occur farther and farther apart. Botanists use the paired groups of leaves to characterize the plant at its various vegetative stage levels. The first pair of leaves are known as the first-order leaves, the next pair are second-order, and so on up to and greater than fourth order, depending on how tall the plant grows. The amount of light controls flowering, so by manipulating the amount of light a cannabis plant receives, one can keep it in the vegetative state for some time. Controlling the amount of light of a plant started from cuttings (cloning) is also a way of circumventing the plant's germination and seedling stage.

The vegetative stage is the phase in which marijuana plants can be "trained." Training forces the plant to grow in a particular direction and also enhances flowering. Some cannabis plants will grow vertically to impressive heights, and growers have realized that to increase flower output it would be better if they grew more horizontally and less vertically. A process called "topping" was therefore developed to encourage plants to grow outward (laterally) rather than vertically.

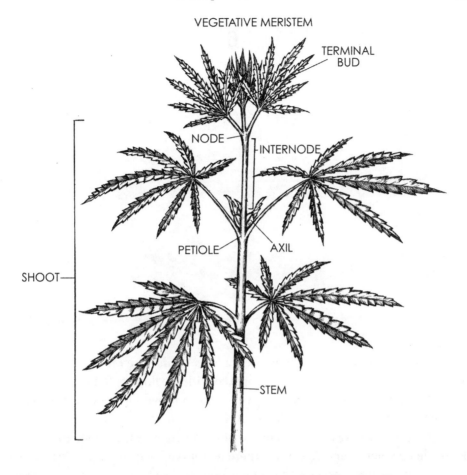

Figure 5.5. Vegetative stage of the cannabis plant, showing the proliferation of leaves.

If the plant is allowed to grow vertically, it will develop a single large set of buds called cola at the top, with smaller, less productive ones on the main stalk. Topping redirects the plant's growth hormones away from the main stalk and to the side branches. Bushier, fuller, but shorter plants are the result.

Although the flowers to be discussed in chapter 6 are the primary production goal of growing marijuana plants, the initial seedling stage and young vegetative stage are critical for establishing the foundation for overall flower production. If the plant is healthy and growing in its early and intermediate vegetative stages, flower production will also be enhanced during the transition to the reproductive phase.

6

Buds

annabis flowers, known in the trade as "buds," are relatively mundane if
not ugly structures when compared with the flowers found on ornamen-
tal orchids, roses, lilies, or even woodland wildflowers such as buttercups or columbines. The homeliness of the cannabis flower is the result of how
the flowers evolved within the cannabis lineage. There was no need to be showy
along their evolutionary trajectory. There was also no need to have colorful petals
to attract pollinators carrying the pollen grains to distant flowers, because cannabis pollen is distributed to distant pistils by the wind. As a wind-pollinated plant,
cannabis's evolution could result in the reduction or even complete loss of petals
and other structures that might block access of the wind to the pollen grains. In
addition, the chemicals produced by cannabis to prevent herbivory are relatively
successful at keeping insects away from the plants (see chapter 1), including the
floral buds. These strategies make for a dull-looking flower, but one that still develops along the same rules as other flowers.

The evolution of cannabis flowers leads us directly to Charles Darwin, one of
the most studied scientists in history. His popularity is due to many factors. Con-
ceiving of natural selection and how it acts as a force for adaptation and evolution
was an amazing accomplishment. Another contributor to his popularity was his
note-taking and his wide sharing of ideas in letters to colleagues. His notebooks

give us an unprecedented record of how the idea of natural selection developed in his mind, a process that is stunning to read. Still another reason is that he could write some of the clearest and most beautiful prose ever composed in science.

An Abominable Mystery

Darwin was one of the most competent zoologists of his time, as his work on barnacles shows. Barnacles were the topic of his primary research—the focus of his "day job." Almost a decade before he published *On the Origin of Species,* he wrote two monographs on barnacles. His interest in them originated while he was sailing with the HMS *Beagle* in 1835. He dedicated two decades to dissecting, examining, and taking notes on them. He also dabbled in plant biology.

Darwin was as much of a science promotor as he was a biologist. He is responsible for such memorable phrases as "this view of life" (on natural selection), "endless forms most beautiful and wonderful" (on biological diversity), "tree of life" (on the relationship of all living things on this planet), "entangled bank" (on evolution), "like confessing a murder" (on a creatorless natural world), and "beautiful contrivances" (on plants). He was astounded by the broad diversity of flowering plants, and he wrote a book on orchids, which was more about natural variation and divergence than about botany: *On the Various Contrivances by Which British and Foreign Orchids Are Fertilised by Insects, and On the Good Effects of Intercrossing.* One section of the book suggests what he was trying to achieve: "Importance of trifling details of structure."

Nature had sculpted small changes in the flowers of orchids, but these "trifling details of structure" have had a huge impact on how orchids interact with pollinators. You need only to look closely at the flowers and the insects that visit them to see that Darwin had a perfect system for studying the effects of natural selection. Most orchids rely on insects for pollination, and the insects in turn rely on orchids for sustenance. While orchids present spectacular and diverse flowers that reflect mechanisms for attracting insects and for presenting them with pollen in a way that will most effectively result in pollination, the insects themselves have undergone special changes that enable them to take advantage of the nectar and pollen rewards that orchids offer. Their relationship has everything one would need to describe natural selection: the orchids are evolving to take advantage of the shapes and behaviors of the insects to effect pollination, and the insects are evolving to best reach the sugary rewards provided by the flowers.

There is, however, a bit of danger in trying to find natural selection in every natural system, as Richard Lewontin and Steven J. Gould pointed out in a wonderful 1972 paper entitled "The Spandrels of San Marco and the Panglossian Paradigm: A Critique of the Adaptationist Programme," known more simply as "Spandrels." Gould and Lewontin made it abundantly clear that not all evolution is caused by selection—historical contingency (you have to use what is passed down to you) and random events, among other factors, are also involved. For orchids, this historical contingency is the basic floral body plan with which they have to work.

Flowers and their sudden appearance in the fossil record were a huge part of what befuddled Darwin. He knew that something spectacular must have happened as plants diverged and eventually evolved the structures we call flowers. The fossil record revealed that a plethora of floral forms appeared abruptly and as soon as flowers first appeared. He could not understand how that much variation in "details of structure" of flowers could occur out of nowhere. In an 1879 letter to his confidant Joseph Hooker he coined the phrase "abominable mystery" to describe the appearance and almost immediate divergence of flowering plants.

There are two ways to establish the timing in divergence of an evolutionary event. Both require fossils. The first method uses the appearance of a fossil to mark its most recent probable origin. The second uses fossils to calibrate change at the molecular level as a sort of clock. If you can measure the divergence of two living organisms at the molecular level with a calibrated "molecular clock," then you can obtain some idea of the date of divergence of those two organisms. Although both strategies sound pretty simple and straightforward, there are some complications.

Darwin had a good sense of deep time, which was necessary for his theory of natural selection to work. He did believe at one point that the earth was 300 million years old. We now know that the earth is more than ten times older (4.5 billion years old) than Darwin's first estimate and that the different geologic eras that Darwin was familiar with are also much older.

Clocks and Gaps

The idea behind a molecular clock is that changes in DNA sequences will accumulate over time in a clocklike fashion, resembling the ticking of radioactive elements used for radioactive dating. Using a molecular clock for plants has almost always pointed toward a Jurassic divergence of the angiosperms. In other words, fossils of plants bearing flowers should appear during the Jurassic based on molecular clock calculations. Most paleobotanists are quick to point out that no fossil evidence exists for Jurassic flowers. But molecular phylogeneticist Jennifer Morris and her colleagues argue that "the sparseness of early land plant megafossils and stratigraphic controls on their distribution make the fossil record an unreliable guide, leaving only the molecular clock," in this case a molecular clock that predicts a Jurassic origin for flowers. Morris's study adds that "the Jurassic angiosperm—essentially a product of molecular phylogenetics—may have become the holy grail of palaeobotany—but it appears equally mythical."

How does a molecular clock work? Consider a gene within an organism's genome. That gene has a relatively stable sequence of DNA; within a lineage, however, there are occasional mutations that occur, and these mutations can be considered like the ticks of a clock. As lineages diverge, mutations (ticks) occur between species in their sequences that can be detected using sequencing technology. Scientists can count the accumulation of ticks and derive an estimate of how many changes have occurred since the divergence of two lineages. As ticks accumulate over a period of time, one can posit an approximate rate of mutation for that gene.

All genes do not tick at the same rate, so some caution should be used when applying a clock. Using fossil data, scientists can correlate the number of ticks with an absolute age and produce some estimate of how old certain lineages of organisms are, based on the number of ticks that have accumulated since their divergence. This molecular clock approach was first suggested by Emile Zuckerkandl and Linus Pauling in the 1960s. Because not all genes evolve or mutate in a regular ticktock-like fashion, this approach remains controversial. However, some genes do appear to be clocklike within some groups of organisms given certain assumptions, and thus can give us a way to convert molecular changes into estimations of the absolute ages of organisms.

Even though recent developments in statistical methods have made estimates from molecular clocks more precise and understandable, the fossil folks and the molecular analysts are almost always at odds. The fossil folks are somewhat touchier about this issue, reflected by this statement from paleobotanist Qiang Fu and colleagues (whom we met in chapter 3): "Studying fossil flowers, especially those from earlier geologic periods, is the only reliable way to get an answer to these questions." And "rather than a *mythical* artifact of genome-based analyses, Jurassic angiosperms are an expectation of our interpretation of the fossil record" (italics my emphasis). Why the tension? Partly because the molecular estimates almost always conflict with the fossil dates, and not only just for plants. The molecular analysts persist in making their estimates in the face of conflicting fossil information. It is understandable that paleobotanists would take exception.

The trend to move on past the fossil data is exemplified by two molecular systematic studies: one by Hong-Tao Li and his colleagues and the other by Mario Coiro and his colleagues. Both studies contended in 2019 that the molecular clock can be used to date the age of flowering plants to somewhere between 140 million and 250 million years ago. From the fossil trenches, a study in 2021 by Daniele Silvestro and colleagues examined a database with over 15,000 angiosperm (flowering plant) fossils, covering almost 200 families (there are about 600 known living families of plants). With that much information, statistical methods can be developed that allow for the estimation of fossil abundance with corrections for gaps in the fossil record. Silvestro and colleagues concluded that "an early, pre-Cretaceous (145 million years ago or more) origin of angiosperms is supported not only by molecular phylogenetic hypotheses but also by an analysis of the fossil record." But yet another study, a 2022 paper by Hervé Sauquet and colleagues, sums it all up: "The age of flowering plants is unknown." By "unknown" they mean that there is so much variance around any estimate that a satisfying divergence date is not possible at this time.

This is the way science works; there is a lot of back and forth and tension on certain issues. Without the tension, science doesn't proceed to more precision. In chapter 3, I noted that we could consider the origin of flowers to be about 150 to 200 million years ago. With the molecular clock included, the conclusion would be that flowers probably arose anywhere between 140 million and 270 million years ago. It means that hundreds of millions of years were involved in the evolution of cannabis flowers. Here indeed is Darwin's abominable mystery in all its glory. No one has convincingly found a Jurassic flower fossil, but it can be strongly inferred that flowers have certainly existed for over 150 million years, an estimate that follows from both molecular clock and paleobotanical studies.

The Age of Cannabis

What about cannabis's origin? As previously mentioned, the divergence time for *Cannabis* from *Humulus,* its closest living relative, has been estimated to have occurred about 28 million years ago using a molecular clock. The earliest fossil evidence for cannabis comes from fossilized pollen found in the Ningxia region of China and is dated at about 20 million years ago. Does the discrepancy sound familiar? Just as with the abominable mystery for angiosperm divergence, the molecular clock estimate for cannabis is deeper or older (30 million years ago) than the earliest documented fossil occurrence (20 million years ago). The discrepancy is not as great as it is for Jurassic angiosperms and could easily be reconciled by further fossil evidence from deposits in the 30-million-years-ago range. There is a Cannabaceae specimen from Mexico in Dominican amber (15–25 million years ago), but it is not a *Cannabis* or *Humulus* species but rather *Aphananthe manchesteri,* a member of a genus not closely related to either hops or cannabis. While it shows that the family Cannabaceae existed in Mexico perhaps 25 million years ago, we cannot say that cannabis or hops were there too. Whether the molecular clock or the paleo estimate wins, it is clear that *Cannabis* (as well as *Humulus*) had between 20 and 30 million years of evolution to get to where it is now.

In doing the research for this book, I came across a popular article claiming that "Fossilized Cannabis Reveals the Plant Is 27.8 Million Years Old." Other headlines took the same sensationalized approach, but this title was particularly galling. First, to my knowledge the oldest fossilized cannabis is only about 20 million years old. Second, that fossil is of pollen, and while it tells us that cannabis ancestors existed then, it doesn't provide any information about the plant itself. Third, when the ancestors of current day *Cannabis* and *Humulus* diverged, they did not immediately form the present-day species *Cannabis sativa* and *Humulus lupus*. Instead they diverged into other ancestors of *Cannabis sativa* and *Humulus lupus,* which eventually evolved into the species we know today.

In fact, there were probably a lot of cannabis species that have appeared over the past 20 to 30 million years. One problem: all but *Cannabis sativa* went extinct. Evolutionary biologist and plant systematist Aelys Humphrey and colleagues have assessed the rate of extinction of plants on our planet. The results are depressing and alarming, because humans have accelerated the rate of extinction for many groups of organisms. Through careful examination of one of the most complete datasets ever assembled to examine plant extinction, they show that the extinction rate for plants today is about 600 times the rate in the prehuman past.

The rate of nonhuman-induced extinction is hard to gauge, but I present here a back-of-the-envelope attempt. An estimate that five families of plants go extinct every 1 million years is a reasonable starting point. Using this family-level rate, we can extrapolate the extinction rate for the genus *Cannabis.* I will use the

Table 6.1. ALL NAMED SPECIES IN THE GENUS *HOMO*

Homo sapiens	†*Homo heidelbergensis*
†*Homo antecessor*	†*Homo longi*
†*Homo erectus*	†*Homo luzonensis*
†*Homo ergaster*	†*Homo naledi*
†*Homo floresiensis*	†*Homo neanderthalensis*
†*Homo habilis*	†*Homo rhodesiensis*
†*Homo rudolfensis*	

Note: Daggers indicate the taxon is extinct.

human lineage to calibrate this rate, despite its limitations. Over the past 2 million years all species of human in the table have been found either fossilized or, in the case of our own species *Homo sapiens,* observed physically. If we assume the list is exhaustive, twelve species of human have gone extinct over the past 2 million years (table 6.1).

Using these numbers, we can calculate that six species per genus per million years would go extinct, suggesting that over 30 million years 120 to 180 species of *Cannabis* would have gone extinct since the divergence of *Cannabis* from *Humulus.* This is probably inaccurate by an order of magnitude too high, so to scale down we might expect there to have been twelve to eighteen species of *Cannabis* that have gone extinct since the genus was formed by splitting from the *Humulus* lineage. Many paleontologists might question my assumptions, but this is more than likely a decent ballpark estimate.

Inner Truth and Logic

One goal of this chapter is to examine the structure and development of cannabis flowers. After all, the flowers are where most of the action in cannabis plants takes place. Thus far, I have taken a rather extensive look at the past history of angiosperm flowers in general and cannabis flowers only fleetingly. This is because I have focused on fossil flowers, and there simply aren't any fossil cannabis flowers (some pollen, but nothing more than that).

Indeed, fossil flowers give us the best idea of what the ancestral flower may have resembled. Understanding the ancestral flower can provide a basic plan on which the diversity of flowers we see today is based. This basic plan is called the archetype, which is an interesting term in and of itself. Two of the eighteenth century's great thinkers—Immanuel Kant and Johann Wolfgang von Goethe—had a lot to do with the development of the archetype concept. Both saw its utility and importance as a way to understand organismic diversification. "The archetypal plant as I see it," Goethe wrote, "will be the most wonderful creation in the whole world, and nature herself will envy me for it. With this model and the key to it, one will be able to invent plants . . . which, even if they do not actually exist, nevertheless might exist and which are not merely picturesque or poetic visions and illusions but have inner truth and logic."

An archetype would provide researchers with a model that could predict the range of plant forms and structures that are possible given the archetype; as such, it is the holy grail for plant developmental biologists. A slew of early nineteenth-century anatomists such as Carl Gustav Carus (Germany), Richard Owen (England), and Étienne Geoffroy Saint-Hilaire (France) furthered the idea. Later in the century Darwin picked it up and developed his well-known ideas about homology.

Goethe, though, had the most interesting and relevant ideas about floral structure, and this is probably because, in addition to being a philosopher, poet, and author, he also focused much of his energy on studying plants. In his search for an archetypal explanation for flowers he was led to conclude that the organs that compose flowers are modified leaves; in his own words, "Alles ist Blat" (All is leaf). It is indeed true that flowers are essentially modified leaves (the leaf archetype idea). Goethe thus invented an interesting starting point for understanding how a flower might develop, and modern models of flower development harken back both metaphorically and literally to his ideas about floral development.

As the following diagram based on Darwin's *On the Various Contrivances* book shows (fig. 6.1), floral structure is basically logical, or rather it can be broken down into logical descriptions. This orderly way of describing floral structures enabled scientists to analyze how flowers develop and to study how their development is controlled by genetic mechanisms. Simply put, flowers consist of whorls of organs (sepals, petals, stamen, and the pistil) that each are considered to have evolved from a leaf. The whorls of organs are concentric, with the sepals being the outermost whorl and the pistil forming the innermost whorl. These organs should be familiar from our discussion of plant sex, because they are all

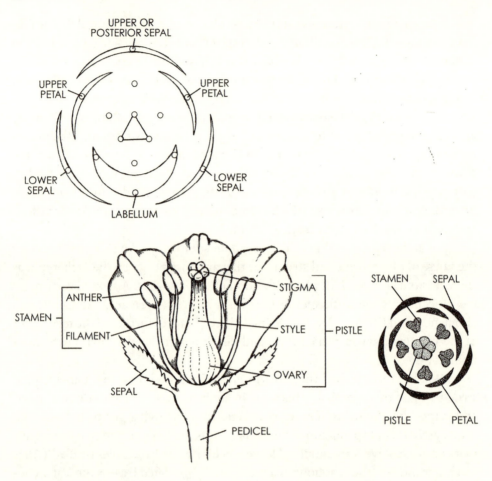

Figure 6.1. Top: Redrawn diagram of the flower from Darwin's *On the Various Contrivances*. Bottom left: Drawing of a typical dioecious flower. This internal side view shows the male and female genitalia (the carpels and pistil form the first whorl of the flower and the stamens the second), as well as the petals that lie in the third whorl around the genitalia and the sepals that lie in the fourth whorl. Bottom right: Top-down view of the whorl structure of a typical flower.

components of plants' reproductive systems. As the outer whorl, sepals can protect the flower when it is in bud. This whorl of sepals is sometimes referred to as the calyx, a term that comes into play for cannabis flowers. Petals, next in the whorl system, are often the showy, attractive organs of the flower. In some flowers such as petunias, the petals fuse together to form a floral tube. The petal whorl is often referred to as the corolla. Stamens produce pollen (which makes the male gametes), and the pistil houses the female gametes inside the ovary. The pistil is formed from a whorl of organs called carpels. While the sepals and the petals can vary enormously across flowering plant lineages and produce beautiful and sometimes bizarre structures, and even the stamens and pistils can also vary broadly, every flower that has ever existed has followed this archetypal structure. Once it was recognized that all flowers followed this archetype, the search was on for explanations and extensions of the archetypal model. With the development of genetic approaches, revealing the secrets of the archetype became increasingly feasible.

Our understanding of genetics has required that scientists focus on what are called model organisms, used to tease apart the genetic mechanisms underlying the development of basic form and function for the lineages they represent. For classical animal genetic systems there are two famous ones: the fruit fly and the nematode. These were chosen because of their rapid life cycles and their ease of rearing and observation in a laboratory setting. Finding a vertebrate model organism was difficult, but *Mus musculus* (the house mouse) was settled on early, and has since developed into a monster of a model system, helping us understand many aspects of human biology and biochemistry. Plants are no different, and the first plant model system was developed using a small weedy plant called *Arabidopsis thaliana* (the mouse-ear thale cress). It was chosen because its generation time is rapid (from seed to flowering takes only around twenty days), and it can easily be grown in a lab on small petri plates with artificial soil in the guise of agar and nutrients. Another plus is that mutants could easily be generated and isolated; the existence of interpretable genetic variation makes for a good model system. Finally, it could easily be observed developing in the lab, and it forms flowers.

Using *Arabidopsis* and eventually several other second-generation plant model systems, researchers set out to decipher the genetic and developmental control of flowers, so that they could realize Goethe's vision of developing a model of floral development that would be "not merely picturesque or poetic visions and illusions but have inner truth and logic." Similar work on cannabis is lacking, and so we need to rely on such models to explain how a cannabis flower might develop.

As Easy as ABC(DE)

Having the archetype in hand and knowing the diversity of floral structure, it would have been possible to elucidate the *mechanics* of floral development in the late 1800s, but we had to wait until 1991 and the development of a model system to elucidate the *genetic mechanisms*. By combining information gained from observations of several important *Arabidopsis* floral mutants, Elliot Meyerowitz and his colleagues at the California Institute of Technology developed a model for the genetics of floral development that quickly became known as the ABC model. The important genetic strains of *Arabidopsis* were discovered over the years as what are called homeotic mutants, in which a full-blown anatomical structure develops in an atypical place. One classic homeotic mutant in fruit flies is called proboscipedia; as its name implies, it is a mutant whose legs bizarrely grow in the position where the proboscis should be—the ultimate foot in mouth. The *Arabidopsis* homeotic mutants were important because they all produced the growth of flower parts (sepals, petals, stamens, and the pistil) in whorls where they didn't belong. As mentioned earlier, the pistil comprises a whorl of carpels; therefore, in the literature for floral development you will see the term "carpels" used instead of the composite organ, the pistil. I will also use carpels as the organs that compose the innermost whorl of the flower.

Mutants were crucial to the ABC story. They told Meyerowitz and his colleagues what factors were present and what these factors did for the flower. For instance, the *Arabidopsis* mutant apetala 1 is a mutant whose sepals and petals were transformed into carpels and stamens, respectively; apetala 3 is a mutant whose petals and stamens were transformed into sepals and carpels, respectively; and agamous 1 is a mutant whose stamens and carpels were transformed into petals and sepals, respectively (see sidebar).

While these mutants are not as bizarre to the untrained eye as a leg growing out of a mouth would be, they do look quite odd to a plant biologist. Single mutants gave the group a lot of information, but they couldn't pin down the model with just single mutants. Luckily, *Arabidopsis* biologists can easily do crosses that result in the combination of two, three, and even more mutants into a single genome. This produces double and triple mutants that were the proof in the pudding for the ABC model. Through these genetic crosses, researchers manipulated the plant to produce all three possible double mutants and the one possible triple mutant. Along with observations of what these combinatorial mutants looked like and their previous knowledge of the single mutants, they were able to develop the ABC model of floral development. In essence, the model suggests that the genetic control of the whorls of the cannabis plant is more than likely controlled by the same mechanisms as in other flowers.

As gene sequencing and molecular analysis became more sophisticated in

THE ABC MODEL OF FLORAL DEVELOPMENT

The ABC model comprises three controlling genetic factors (fig. 6.2). These could be gene products or regulatory factors or even more complex developmental factors, but Meyerowitz and colleagues demonstrated that in flowers there are three: A, B, and C. The presence of these three factors across the whorls of the developing flower dictates what organs will develop in each whorl. As figure 6.2 shows, if the tissue has only the A factor, then the organs will develop into sepals. Likewise, if only factor C is present, then carpels will develop. To eventually become petal tissue, an organ primordium (the tissue that will form the mature organ) needs to have both A and B factors, and to become a stamen it needs to have both B and C factors. One other variable is that two of the factors are antagonistic to each other. That is, if one is present, the other won't be expressed in that spot. And if one is absent, then the other fills in.

The A and C fields were found to be antagonistic fields; thus, if A was absent as in the apetala 2 (ap2) mutant, then C is produced in all four fields and B is still produced in fields 2 and 3. Thus, the ap2 mutant flower has a pattern of C, B + C, B + C, C that gives rise to a flower with the following organs in its four whorls: carpels,

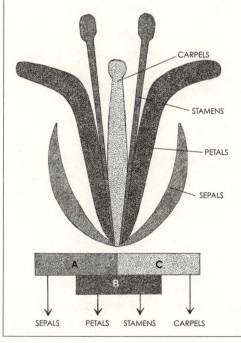

Figure 6.2. The ABC model. There are four developmental fields corresponding to the four structural whorls in a flower: carpels, stamens, petals, and sepals. There are three controlling factors (A, B, and C), and the identity of the four whorls requires the presence of the controlling factors as shown at the bottom of the figure. Adapted from Chanderbali et al. (2016).

Continued on next page

stamens, stamens, carpels—an odd-looking flower, but one that follows the ABC rules. Likewise, if C was absent, as in the agamous 1 (ag1) mutant, a flower would be made of sepals, petals, petals, and finally sepals in the center. This mutant can be explained as lacking the C function. Again, because A and C are antagonistic, when there is no C function present, A fills in to produce the four fields: where there is only A, which dictates a sepal; A and B, which dictate a petal; then another field with only A and B, which dictates a petal; and finally a whorl with only the A factor, which dictates a sepal for an overall sepal-petal-petal-sepal structure (fig. 6.3). Meyerowitz and colleagues were able to explain all the interesting forms produced by single, double, and triple mutants with this model. Concurrently with the Meyerowitz group's work on *Arabidopsis* in California, a British group led by plant molecular geneticist Enrico Coen was working on snapdragons (*Antirrhinum majus*), and they discovered the same basic model. This meant that the overall ABC model was likely conserved across a wide range of flowering plants, including cannabis.

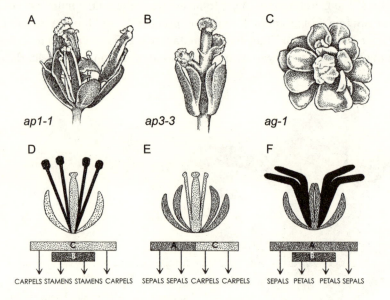

Figure 6.3. The three kinds of mutants used by Meyerowitz and colleagues to elucidate the ABC model. A: apetala1 (ap; there is another mutant in this class that Meyerowitz and colleagues used called apetala 2 or ap2). B: apetala 3 (ap3; there is another mutant in this class that Meyerowitz and colleagues used called pistillata 1 or pi1). C: agamous (ag). D, E, and F show the gene expression patterns underlying the floral mutants corresponding to A, B, and C, respectively. Adapted from Alvarez-Buylla et al. (2010).

Continued on next page

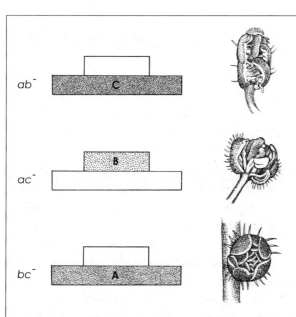

Figure 6.4. Diagram of double mutant ABC model factors, and drawings of *Arabidopsis* floral mutations corresponding to the double mutations. Adapted from Alvarez-Buylla et al. (2010).

What do you think would happen when you have a double mutant that wipes out both factors A and B? When this double mutant develops, it will permit only the C factor in all whorls. And the logic of the model tells us that having only the C factor means a carpel identity in all four whorls: carpel-carpel-carpel-carpel (fig. 6.4).

What about a double mutant missing the B and C factors? Again, having only factor A will dictate a sepal identity in all four whorls: sepal-sepal-sepal-sepal. The double mutant that is missing A and C factors is difficult to predict but also quite logical. It is difficult because there is no simple identity predictable for a field that is making just the B factor. When factor A and B are together, a petal results. And when factor C and B are together, a stamen results. This means that B allows the field to choose between being a petal (A + B) or a stamen (B + C). When only B is there, the identity is both stamen *and* petal.

But only the two middle whorls will be both stamen and petal. What about the outer and inner whorls? When there is no factor made in a whorl, what happens? This is where Goethe comes back into the picture. Remember that he suggested that the primordial state of a flower would be a leaf. In essence, when you have a triple mutant that lacks factors A, B, *and* C, you will get four whorls of leaves. We still haven't explained the double A/C mutant that makes only the B factor, but now we can. The outer whorl will be one of leaves as well as the inner whorl, making the flower structure as follows: leaf – (petal + stamen) – (petal + stamen) – leaf.

the later 1990s, the gene products of the three factors were quickly deciphered. Not surprisingly, it was discovered that the genes that code for the A, B, and C functions are what are called regulatory genes or, more precisely, transcription factor genes. These genes act like valves that facilitate the turning on and off of the gene product. We know they are transcription factors because their DNA sequences include short motifs of conserved nucleotides that, when translated into protein, bind with other nucleic acids to regulate the target gene's expression. All three factors are genetically encoded by what are called MADS box genes. The name "MADS box" comes from a monstrous combination of mutant names from four very different organisms, but their structure at the genetic level is quite simple and elegant. Another group of transcription factors involved are called EREBP (ethylene-responsive element binding proteins) transcription factors. Both MADS boxes and EREBP transcription factor genes are gene families, meaning that there are many members of that kind of gene in a genome. The existence of many kinds of MADS boxes and EREBP factors indicates that these genes are more than likely involved in generating the huge amount of anatomical variation in flowers.

The logic of the ABC model is both simple and stunning. But like anything in nature, if you look hard enough you will find something that is an exception, or many things that are exceptions. Meyerowitz and colleagues explained this phenomenon in 1991, when sequencing genes and studies to elucidate the molecular nature of genetic systems were still in their infancy. The patterns got even more complicated when flowers that aren't part of the known model systems were examined. Novel patterns of floral structure in these other kinds of flowers led to the need for more explanation, and like any good model the ABC model could accommodate additional factors to explain novel variation. In its current form, the more inclusive model is called the ABCDE model, which includes two more controlling factors, D and E. The D and E factors are important for telling the floral primordium to recognize that it will sit as the terminal structure on a stem. Because the role of the D factor is species-dependent, the general model has been shortened and is known as the ABCE model.

The need for yet another model is suggested because there are structures other than sepal, petal, carpel, and stamen that make up flowers. The best model would not only be able to explain floral development in *Arabidopsis* but also in other plants. The ABC model accommodates four developmental fields (the four whorls), but what if the definition of a flower is everything that is interior to and including a sepal? This would include the ovules of the female part of the flower, which would require an explanation for five developmental fields. And what about flowers that have strayed from the typical sepal–petal–stamen–carpel structure? Indeed, some flowers do away with their petals. They do not eliminate that outer whorl but replace it with something else. This is where flowers such as lilies are relevant; they have modified their two outer whorls into what are called tepals. Tepals share characteristics of both sepals and petals and are found throughout the monocots—such as lilies and tulips. Lilies have the flower structure of tepal–tepal–stamen–carpel.

As Goethe wrote, "With this model and the key to it, one will be able to invent plants . . . which, even if they do not actually exist, nevertheless might exist." Too bad he didn't make a deal with the devil to live to the end of the twentieth century. He would have been quite impressed with himself.

2000 (Floral Development)

In chapter 5 we followed a cannabis plant through the 0000 and 1000 development stages (according to Mediavilla and colleagues' digital scheme). It is now time to take it through the 2000 stage: the part of the plant's development during which flowers form. This stage is complicated, because cannabis has two sex systems (monoecy and dioecy), which is why the 2000 level of digital development is subdivided into 2100 (dioecious male, or a plant that makes only male flowers), 2200 (dioecious female, or a plant that makes only female flowers), and 2300 (monoecious, or a single plant that makes both male and female flowers). In addition, some cannabis plants produce flowers that are hermaphroditic, or have both male and female organs in the flower (fig. 6.5). By far the most instructive way to examine the formation of flowers of one species is by comparing them to other closely related species. Fortunately, studies comparing cannabis to its close relatives exist. Members of the genera *Trema* and *Celtis* have been used to understand floral development in the Cannabaceae, and this has given researchers a good idea of the intricacies of floral development in the family in general and cannabis specifically.

The floral morphology of cannabis has a whorled architecture, just as discussed for *Arabidopsis*. Since the flowers are mostly unisexual, however, we need to discern between male (staminate) and female (pistillate) flowers. Because cannabis flowers have lost their petals, they only have sepals as their nonreproductive floral organs. The sepals are green and leaflike and protect the flower when in bud. As detailed in chapter 5, the onset of floral development is highly dependent

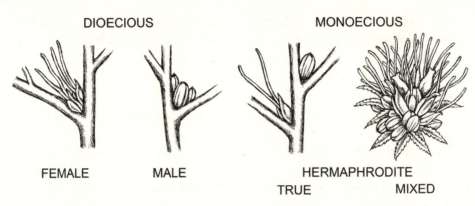

DIOECIOUS   MONOECIOUS

FEMALE   MALE   HERMAPHRODITE
                                     TRUE   MIXED

Figure 6.5. Flower proliferation in female, male, true hermaphrodite, and mixed-gender cannabis plants (from left to right). The flowers grow in the axial areas of diverging branches.

on the photoperiod that the plant experiences. Once the photoperiod is optimum, flowers will begin to develop. The left side of figure 6.6 illustrates a staminate (male) flower.

Note the composition of the male flower structure. The black crescent at the bottom is a bract, which is a modified leaf that subtends the flower itself. In terms of the ABCE model, the bract is simply a modified leaf that is "outside" and not part of the model. The light gray crescents on the right and left are called prophylls and may be what is considered the first field, while the medium gray crescents are sepals (together they are called the calyx), making up the second developmental field. This whorl is followed by the whorl of stamens. Since this is a diagram of a male flower, there is no whorl of carpels at the center.

The sequence of events of the development of the staminate (male) flower is as follows: The meristem of the developing flower is rounded; all the subsequent floral structures will emanate from this meristem as the result of genetic signals and protein interactions. The first tissues that will develop into sepals appear in a spiral whorl, and the sepals begin to develop. The tissue that will eventually become the stamens appears and develops in a spiral pattern too. At the same time the sepals elongate. The stamens take up the central position of the flower, excluding the development of carpels and ovule and completing the components of the male flower. Important structures known as trichomes arise on the upper and lower sides of the bract and the calyx as the flower develops. Chap-

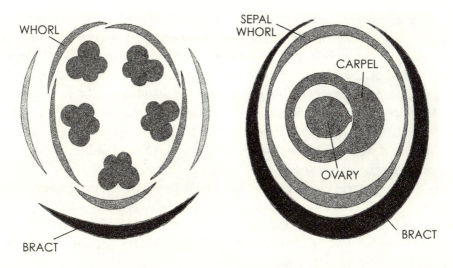

Figure 6.6. Left: Diagram of a staminate cannabis flower. The sepals together form the calyx. Right: Diagram of a pistillate cannabis flower. Adapted from Leme et al. (2020).

ter 9 discusses the trichomes in more detail, because these tissues are important for producing the thousand or so compounds made by the cannabis plant.

The pistillate (female) cannabis flower also has a subtending bract (fig. 6.6, right side), but it grows laterally to nearly encircle the entire inner developmental fields. The female flower lacks prophylls but has a fused calyx (or sepal whorl) that encloses the carpels. Two carpels fuse to form the ovary. The sequence of events of the development of the pistillate cannabis flower is as follows: The early floral meristem is dome-shaped and protected by a large bract. The first sepal primordium appears, followed closely by the appearance of the second sepal primordium, which appears to fuse with the first to form a ring of sepal tissue around the base of the flower. At the center of the still developing flower, the tissue that will become carpel elongates, and two carpels form and fuse to make up the ovule. The carpels close at their tops, and the tips of the carpels elongate to form the styles (the tubes that allow pollen access to the ovules). The calyx ultimately forms as a cuplike structure that covers the ovary. The subtending bract will develop large numbers of trichomes and enlarge to eventually enclose the entire female flower.

It appears that cannabis flowers have the typical four- or five-whorled structure of most flowers and could therefore fit the ABCE model nicely. However, in a 2020 review of cannabis genetics and development, plant geneticist Gianni Barcaccia and colleagues pointed out a lack of information on how the ABCE model applies to floral development in the *Cannabis* genus. This is more than likely due to the lack of formal work on the genetics of *Cannabis* because of now-removed legal restrictions and the difficulties of performing genetic experiments with cannabis plants. Their generation times are long, and they are rather big plants (unlike the tiny *Arabidopsis* plants that can be grown on a petri dish). But Barcaccia and his colleagues thought of a different way to extend the ABCE model to cannabis. To explore whether the genes involved in floral development might fit the ABCE model for cannabis, the group scanned cannabis genome sequences to see if they could identify some of the same genes that are involved in the floral development in *Arabidopsis*. They found almost all of the MADS box transcription factors that facilitate the ABCE model in the cannabis genome.

The structures of the cannabis flower have clear affinities or homologies with other flowers. It is also clear that cannabis has most of the genetic toolbox to participate in an ABCE-like developmental dance. It is likely that *Cannabis* uses a similar genetic toolkit to make its flowers, with modifications to the ABCE model that specifically address traits inherent to cannabis, such as the loss of petals and the formation of unisexual flowers. Further genetic research will, as Goethe wrote, reveal the "inner truth and logic" of the cannabis flower—perhaps both aesthetically and commercially.

7

Decarboxylation

Cannabis ingestion has historical and cultural contexts that are part of the cannabis story. This book does not endorse the use of cannabis, and the discussion that follows in this chapter does not advocate cannabis ingestion. It is well known that many health issues and problems can arise. All cannabis users should be aware that vaping, the use of water pipes or vaporizers, and other methods of consuming cannabis can be harmful to human health. Furthermore, some methods of ingestion, such as shared use of water pipes or other devices, can contribute to the spread of infectious diseases and should be avoided.

The first beings who ingested cannabis probably ate it. And the chemicals that the cannabis plant synthesizes are many and varied. The two that are most relevant to its recreational and medicinal use are tetrahydrocannabinolic acid (THCA) and cannabidiolic acid (CBDA). It is their derivative activated chemicals cannabidiol and tetrahydrocannabinol (THC and CBD) that we need to ingest for cannabis to work its wonders. In chapter 8 we will follow a THCA and a CBDA molecule synthesized by a *Cannabis* plant on their journeys into and through the human body. But before taking that journey, we need a little chemistry and then a bit of head-shop madness to get the THC and CBD in proper form for their effects to work.

Light Up or Leave Me Alone

Figure 7.1 depicts the chemical structures for THCA and CBDA. There is one small part of these molecules (called a carboxyl group) on which we need to focus at this juncture. The rest of the structure is discussed in chapter 9.

To simplify these structures, I have placed a "black box" over most of both molecules in figure 7.1. This black box is the same in both THCA and CBDA, so not seeing what is obscured doesn't matter. THCA and CBDA are inactive in their acid forms, but by altering them to remove the carboxyl group (COOH), we get active THC and CBD. Note that the carboxyl group is in the same place in both CBDA and THCA. This similarity suggests that we can devise a way to alter this part of both molecules to release the COOH.

Carboxyl groups are found all over the organic and biochemical world. One of the most common kinds of carboxyl groups are those in amino acids, which have a carboxyl end (COOH) and an amino end (NH_3). These two kinds of ends in amino acids are highly reactive, which is why amino acids can be strung together in long chains to make proteins. The carboxyl and amino ends create strong bonds with each other.

Carboxyl groups have many functions in nature. For instance, by simply adding a hydrogen component (H) to a carboxyl group we would get formic acid, a compound made by several species of ant as a defense mechanism. By releasing formic acid, which is noxious to other animals, the ant can fend off predators.

So how do we remove the carboxyl group to activate the THCA and CBDA? The answer is complex and, of course, chemical. We need to yank the COOH group away from the black boxes for both THCA and CBDA, which in chemical lingo means to decarboxylate the THCA and CBDA. If we call our black box and the substance coming off the lower left of the THCA black box R_T, then THCA can be simplified to R_T-COOH. If we do the same for CBDA (that is, call the black box and the substance coming off the lower left of the CBDA black box R_C), then CBDA can be simplified to R_C-COOH. The reactions are simply:

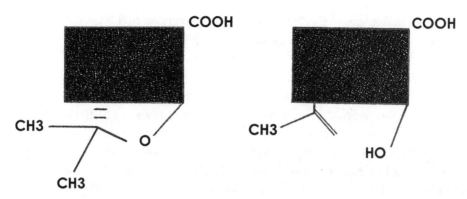

Figure 7.1. Chemical structure of THCA (left) and CBDA (right), the two major psychoactive cannabinoids made by *Cannabis*. The black boxes simplify the chemical structure, as it is the same in both THCA and CBDA. The carboxyl groups (COOH) come off the black boxes in the upper right-hand corners.

For generating THC = R_T-COOH → R_T-H + CO_2
For generating CBD = R_C-COOH → R_C-H + CO_2

These reactions lead to the active cannabinoids (R_T-H = THC, and R_C-H = CBD) and carbon dioxide (CO_2). There are, however, many ways to decarboxylate, as the names of different decarboxylating reactions indicate: Barton decarboxylation, Kolbe electrolysis, Kochi reaction, Krapcho decarboxylation, Tsuji—Trost reaction, and Hunsdiecker reaction. There are also many different enzymes that can decarboxylate the carboxyl groups. Which one would transform THCA and CBDA into THC and CBD, respectively? None of the ones mentioned above! Instead, heating the plant material (the buds) is the desired way to decarboxylate when *Cannabis* is ingested. But it is tricky: too little heat will under-decarboxylate, and too much heat will destroy the active THC and CBD.

The dynamics of heating cannabinoids have been studied by many chemists looking for that sweet spot where there is maximal THC and CBD activation and minimal breakdown to molecules other than THC and CBD. The technique of high-pressure liquid chromatography (HPLC) is used to quantify THC activation, which is simply the ratio of the amount of THC after activation to the amount of THCA and THC before activation. In inactivated *Cannabis* buds there is a little bit of activated THC because of the drying out of buds. But that is not enough to produce a psychoactive impact, and so chemist Franz E. Dussy and colleagues attempted to determine the "sweet spot" for conversion of THCA to THC using a typical titration experiment. First, they took a dried bud, lopped off a small piece, and used HPLC to measure the amount of THCA and THC in the original dried buds. They next heated the remaining cannabis bud to 120 degrees C (248 degrees F) and then took another part of the bud for HPLC analysis for THC and THCA. The next step involved raising the temperature to 140 degrees C (284 degrees F), and again they took a part of the bud and measured the THC and THCA content. They kept increasing the temperature up to 180 degrees C (365 degrees F). If heat had nothing to do with the conversion, there should be the same ratio of THCA to THC at each step of the experiment.

At 120 degrees C (248 degrees F), the conversion of THCA to THC was only 20 percent, but by 140 degrees C (284 degrees F), there was only a bit of the THCA left in the bud material. By 160 degrees C (320 degrees F), all the THCA had been decarboxylated, but two new molecules started to appear along with the activated THC. These two molecules were cannabinol (CBN) and dihydroxy cannabinol (dihydroxy-CBN). Both byproducts are not psychoactive and are for the most part unwanted when THCA is decarboxylated. Heating to 180 degrees C (365 degrees F) merely increased the amount of CBN and dihydroxy-CBN at the expense of activated THC.

The sweet spot thus lies somewhere between 140 and 160 degrees C (284 and 320 degrees F), but probably closer to 160 degrees C (320 degrees F). When someone lights a marijuana cigarette with a burning match or a lighter, the cannabis that is exposed to the fire burns at about 260 to 370 degrees C (500 to 700 degrees F). This flash of heat will cause some of the THCA in the fire exposed part of the cigarette to be lost; in fact, a good portion of the cannabinoids in that part of the cigarette will be shot, due to the high temperature. But as one smokes the burning cannabis from a marijuana cigarette, the air circulating through the burning material cools it down, so the sweet spot is only roughly attained when smoking buds via rolled cigarettes.

The vapor points of THCA and CBDA are different; if we focus on obtaining THC, then that temperature is not optimal for CBD. In fact, the vapor point of THCA is about 8 degrees C (14 degrees F) lower than for CBDA. Equally important are the vapor points of THC and CBD, which are the temperatures at which these cannabinoids are converted to nonactive forms. The difference is 10 degrees C (18 degrees F) lower for THC than CBD. These temperature differences are the result of the different molecular structure of the acids (THCA and CBDA) and the cannabinoids (THC and CBD). By manipulating the decarboxylation temperatures one uses, different concentrations of cannabinoids can be produced. Length of heating time is also involved, and lower temperatures for longer times will do the same trick as higher temperatures for shorter times. Basically, if you heat the buds near the vapor temperature of THCA (at about 157 degrees C, or 315 degrees F), you will convert THCA to THC, and the CBDA will be left largely unchanged. This protocol will give you more of a psychoactive effect than a medicinal or body effect. Raising the cooking temperature to 180 degrees C (355 degrees F) eliminates most but not all THC and optimizes the concentration of CBD, delivering more of a body effect but a reduced psychoactive one. Raising the bake temperature to 200 degrees C will eliminate the THC and convert more of the CBDA to CBD, resulting in an almost complete body effect.

Pipes, Bongs, and Percolators

"Rolled cigarettes" sounds a bit nerdy and technical, so let's add some color to the discussion by calling it what it is: a joint, a bomb, a reefer, a spliff, a bifter, a blunt, a bomber, a stick, a zol, a doobie, a doob, or a toke tube. But even with so many names, there are better ways to deliver cannabis smoke to the lungs than via a spliff. These means involve pipes of some sort. There are two reasons to go to pipes for inhaling reefer; first, to cool the burning buds so that the decarboxylation of THCA and CBDA to THC and CBD is better optimized, and second, to soften the harshness of the smoke.

Smoking pipes is an old practice. As with any long-established human behavior, it is difficult to trace its origins. Some pipe experts suggest that copper pipe use occurred as early as 2000 BCE in ancient Egypt. But the evidence for this conclusion comes from hieroglyphs, and there is some controversy regarding whether the pipes were used in religious ceremonies or as implements for smoking. If researchers want to claim something is a smoking pipe, then they had better prove that there was tobacco or some other compound in the pipe. As discussed in chapter 2, burnt plant material such as tobacco or cannabis does leave an identifiable resin where the burning occurred. A paper published in 2018 by paleoethnobotanist Stephen Carmody and colleagues announced the discovery of the New World's oldest pipe from a Native American archaeological site in Alabama. The animal bones lying near the limestone-carved smoking tube were radiocarbon-dated to between 1685 and 1530 BCE—about 3,700 years ago. Carmody and colleagues were also able to examine the resin in the so-called medicine tube and found that it was dominated by tobacco remnants. This discovery pushed the oldest known pipe back 1,000 years earlier than what was known previously and has since been considered the oldest known pipe.

According to John Edward Philips, a pipe expert, African pipes date back at least to the sixth century AD. Since this time was well before the introduction of tobacco to Africa, it can be assumed that the users of these pipes were not smoking tobacco. There is some controversy over what was being smoked in these pipes, and one researcher (Jean-Paul Lebeuf) suggests that the plant material was from *Datura metel,* a close relative to the North American jimson weed (*Datura stramonium*), which is poisonous. Philips believes that it was more than likely cannabis that was being smoked in these 1,500-year-old pipes.

One of the most bizarre African pipes is a more modern one called an earth pipe. It starts with a container of some sort, usually a bottle buried in the ground. A tube extends from the bottle and serves as the mouthpiece. Lighted embers are placed in the bottle, marijuana is loaded into the bottle, and the resulting smoke is inhaled through the mouth tube. The earth surrounding the tube in the ground serves to cool the smoke as it is inhaled from the buried bottle (fig. 7.2).

Other experts focus on the elegant pipe designs that were created after the introduction of tobacco to Europe from the New World. Water pipes have a relatively short history. Also known as bongs, they are a smoking device that forces the drawn smoke through water to both control the temperature for efficient decarboxylation and to cool the smoke for the lungs. The word "bong" comes from the Thai word *baung,* which is a bamboo water pipe, but Thailand is not where the bong originated. Hookahs (also called narghile, hubble-bubble, goza, argileh, and shisha) are other good models for how water pipes work, but they are not the earliest ancestors. These water pipes were used to smoke flavored tobacco; they were introduced in India and the Middle East about 500 years ago and later refined in Turkey.

The origin of the bong probably dates back to Africa at about 1400 BCE, based on the discovery of bong-like apparatuses in Ethiopia. Again, since this period was well before the introduction of tobacco to Africa, these water pipes were probably used for cannabis. There is a huge gap in the history of water pipes between the African archaeological pipes (1400 BCE) and the first hookahs (AD 1500). A more recent revolution in water pipes involved the use of glass and plastic in the making of the pipes. This golden age of bongs also coincided with the expansion of recreational use of cannabis, especially from the 1960s on.

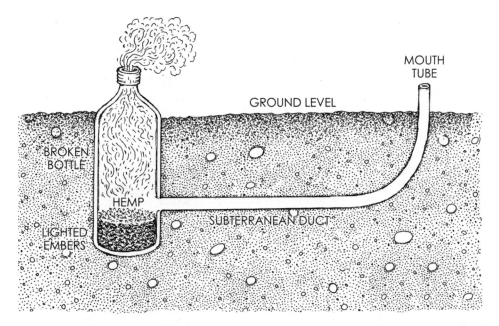

Figure 7.2. African earth pipe. The subterranean duct cools the marijuana smoke, as does the submersion of the lighting chamber buried in the ground. Adapted from Philips (1983).

Gravity bongs and waterfall bongs can be constructed from simple materials (such as used soda bottles and buckets), as shown in figure 7.3. Both devices operate on the idea of controlling the pressure on the lit cannabis without the user dragging on it. Gravity and suction pressure do the work of getting the smoke from the lit bowl. The result in both designs is to cool the smoke being drawn into the chamber and to disperse it a bit. In both designs the user simply inhales the cooled smoke from the top of the device after it has been drawn into the chamber.

There are thousands of makes and models of bongs; a well-established head shop may carry hundreds. To the untrained eye they look aesthetically pleasing, but the shapes and devices in bongs determine the smoking experience. The basic principle of a bong is simple (fig. 7.4).

A bong can be accessorized by several add-ons. Ice can be added to the bong water, but because it can easily move around a pinch is introduced in the chamber to catch or prevent the ice from going too high. Adding ice to the bong cools the smoke coming through from the bowl and downstem. An ash catcher can be

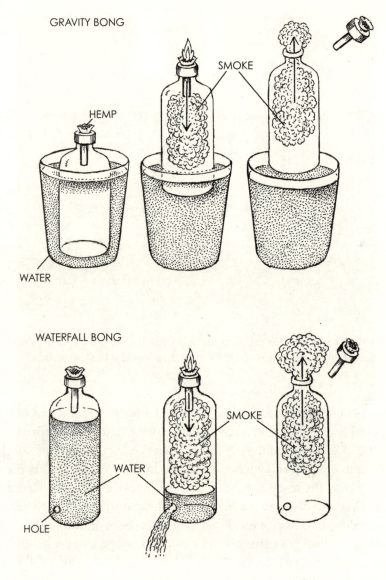

Figure 7.3. Gravity bong on the top, and waterfall bong on the bottom.

attached to the downstem to prevent particulate matter from entering the water. Diffusers are alterations to the downstem that further cool the smoke coming into the chamber: small holes are placed along the downstem to allow smoke to escape at many points in the bong water. These alterations will increase the contact

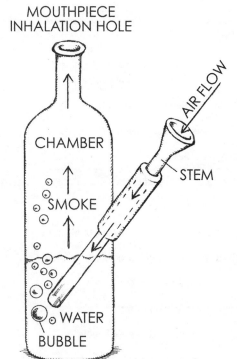

MOUTHPIECE
INHALATION HOLE

CHAMBER

SMOKE

AIR FLOW

STEM

WATER

BUBBLE

Figure 7.4. A lateral view of a typical bong. The flask itself can be of any shape that can hold a reasonable layer of water. A bowl is connected to a pipestem, with the nonbowl end submerged into the water. Some bongs have a small release hole above the water level called a carb or a shotgun hole. When this hole is blocked (usually by a thumb or finger), the smoke will well up in the stem. Another way to deliver the smoke to the user without a shotgun hole is to build in a removable downstem that when removed releases the smoke in the chamber by suction pressure. When the downstem is in place, the system is sealed; but when removed, the user's inhaling takes over, and the smoke pours out of the mouthpiece into the user.

of the smoke with the water, which helps to cool the smoke and make it thicker and smoother to inhale, but they can also increase drag (the amount of effort the user has to put into inhaling).

There are more complex percolator devices (commonly called percs) that can be added to the stem of a bong to cool, smooth, and expand the smoke being inhaled into the bong. Tree percs are the most common, and basic kinds of perc can have a series of vertically arranged tubes through which the smoke is drawn and then released to the bong water through slits that allow smoke to percolate through the water. Table 7.1 presents a list of such percs and what they can do.

Ultimately, bong technology has progressed to the point where the smoke making its way to the lungs of the user is more concentrated with THC and CBD and physically smoother than smoke inhaled from a regular pipe or a spliff, making the sweet spot even easier to maintain. But perhaps the best way to inhale cannabis is by vaporizing, or vaping. Vaporizers can maintain a steadier and more appropriate temperature of inhalant than spliffs, pipes, or bongs.

Table 7.1. TYPE OF PERCOLATORS (PERCS)

Kinds of percolators or "percs"	Comment
Tree	One to eight vertical "arms"; slits in arms; increased flow paths to water
Dome	Large single dome; smoke forced out of dome via hole in bottom
Dewar	A specialized dome perc; eliminates back pressure
Coil	Similar to a tree perc, but arms coiled
Ring	Also known as halo or disc perc; round glass tube
Pinwheel	Extremely rare; a variation on the tree perc
Inline	A combination of tree and dewar perc characteristics, but in series
Honeycomb	Submerged horizontal disc with multiple holes like a honeycomb
Fritted disc	Highest producer of bubbles; similar to honeycomb
Tornado/cyclone	Produces a tornado/cyclone effect in bong water
Showerhead	Acts like a showerhead, but pointing up to release bubbles
Filter disc	An extreme kind of honeycomb perc; superfine holes and several discs

I previously explained that THCA and CBDA decarboxylate by heating to about 160 degrees C (320 degrees F). Part of the problem with all the approaches to creating smoke for inhalation is that they require fire that is much hotter. If one can heat marijuana up to just 160 degrees C (320 degrees F) without burning it, then the loss of THC and CBD will be minimized. THCA and CBDA will vaporize around this temperature, and the vapor can be collected and then inhaled. The easiest way to do this is with a commercial vaping device, but to the casual user they are black boxes. Figure 7.5 shows a diagram of a homemade vaporizer using a heat gun. Other homemade vaporizers can be made from light bulbs (rather like the original Easy-Bake Ovens made for children) or by using a candle and indirectly heating the cannabis in a metal tray to the vapor point. A vaping pen is simply a miniaturized heat gun and collection device.

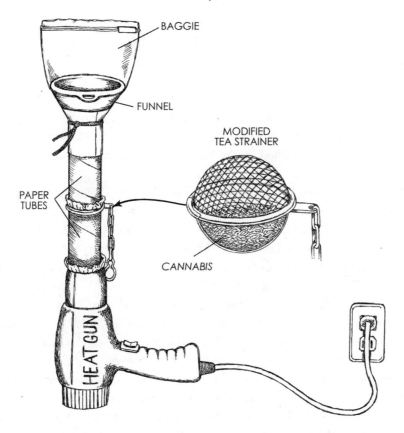

Figure 7.5. A homemade vaporizing device. A holder for the cannabis is constructed with a small glass bowl and a mesh cap (to allow the vaped cannabis
to escape the bowl) between two short columns. A collection bag is placed at
the top of the device and a heat gun at the bottom. The heat gun is then turned
on to force heat up to the glass bowl. The bowl is slowly heated without any
flame touching the cannabis. The heat gun is kept at a steady temperature to
force decarboxylation, but not breakdown to CBN and other unwanted cannabinoid byproducts.

Eating Cannabis

We eat many plants for nutrition, and some of us eat plants for molecules that
have psychoactive effects. Two specific examples should come to mind: chewing
tobacco and "chewing" coca (cocaine) leaves. As with cannabis, the most efficient
way of delivering nicotine (the active chemical we desire in tobacco) is through

smoking it. But some prefer to chew it. When it is chewed, the salivary proteins of the mouth—specifically one called salivary amylase—digest the materials in the tobacco leaf. Another fluid secreted by the mouth called mucin lubricates the chewed matter. When a tobacco plug or a coca leaf is chewed, therefore, the leaf is partially digested, releasing any chemical that was produced by the plant; this includes nicotine. Some chewers swallow the salivary juice produced when chewing tobacco. Others find such swallowing quite unpleasant. Either way, the saliva with nicotine in it is absorbed by the body, and the nicotine eventually is delivered to the brain.

"Chewing" coca leaves is another matter, though. This way of ingesting the psychoactive compound of these leaves—cocaine—is endemic to the native peoples of the Andes. In fact, La Paz, Bolivia, houses the Museo de la Coca, a museum dedicated to the understanding of cocaine, the plant that produces coca, and the sociocultural aspects of coca use by native peoples. It is well known that the people of the Andes used and continue to use coca leaves to alleviate pain, headaches, stomach ailments, cramps, and a slew of other sources of human pain. Calling this "chewing" coca leaves is actually inaccurate, however, because the native people who ingest cocaine orally are really sucking on the leaves. The practice is old, certainly hundreds of years. The veins of a coca leaf are tough and can irritate the mouth, so they are usually carefully removed before placing the leaves in the mouth. The leaves are massaged in the mouth and bathed with saliva, and the person "chewing" will gently try to massage the leaves with the inside of the mouth, just enough to disrupt the leaves' cell membranes but not enough to macerate them.

About ten grams are gently digested like this at a time, and often a bolus of leaves is left in the mouth between the gums and the cheek. The natives know the precise spot in the mouth where saliva is emitted (called the parotid salivary gland), and so the bolus can sit there over this gland. The plant material is left in the mouth for ten to fifteen minutes when the user adds a solution called *llijta,* which is alkaline in pH. *Llijta* consists of vegetal ashes from quinoa or plantain plants. A more modern method is to simply ingest some baking soda (sodium bicarbonate) with the leaves. Such alkaline solutions added to the chewed coca leaves facilitate the activation of alkaloids. Cocaine, the targeted substance from chewing coca leaves, is a tropane alkaloid, and it too is activated. Shortly after the administration of the *llijta* or baking soda, the cocaine kicks in and the mouth, cheek, tongue, and throat numb up as the drug is absorbed into those tissues. The juice is also sometimes swallowed, which can be useful for alleviating pain in the digestive tract. The cocaine absorbed via the mouth and or the digestive tract enters the blood system and is transported to the rest of the body.

Chewing coca leaves is not the most efficient way to get cocaine into the body. It shouldn't be too surprising to many readers that the most common way of getting cocaine into the blood system is to snort it. The snorted cocaine gets into the nasal passageway, where the drug can enter the blood system via a fine network of blood vessels.

Can you snort marijuana? Yes, but any marijuana that is snorted needs to be decarboxylated and powdered, so that the fragments coming into the nose are not so big as to irritate the nasal passages. A California company created a powder product called Canna Bumps. Apparently, the white powder product and name (bumps) conjured up snorting cocaine, a huge marketing mistake. Most marijuana users are not cocaine users, so consumer demand was very low, and the company dropped the product quickly. More palatable to users are nasal mists. These products infuse THC and CBD into nasal sprays.

The nasal passage is an interesting way to ingest cannabinoids, but most users of marijuana ingest it through foodstuffs. Cooking cannabis-laced food items requires a decarboxylation step before infusing the cannabis material into a substrate such as butter, and usually consists of simply controlled baking of the spread-out buds in an oven. Another method is to place the buds in a plastic bag and then immerse the bag in a precisely calibrated pot of boiling water, so the decarboxylation temperature is controlled. Still another way to decarboxylate is to use a precisely calibrated crock pot. Some kinds of butter are better at maintaining and delivering active THC than others. Cocoa butter is preferred over normal dairy butter, and coconut oil over both. This is because coconut oil has more saturated fats in it than either butter or cocoa butter, and these fats assist in the degradation of cannabis glandular material. More important is that coconut oil has a high concentration of medium-chain triglycerides, which accelerate the absorption of the THC and CBD into the bloodstream. It is critical that this absorption occurs in the portal vein and liver, where the active form of THC is converted to the much more powerful and psychoactive 11-OH-THC that eventually will make it to the brain.

Purveyors of edible marijuana have been clever in how they make their products, and the practice is thousands of years old. According to Ram Nath Chopra, considered the "father of Indian pharmacology," cannabis preparation is known from the Atharva Veda, a Hindu scriptural text of procedures for everyday life. The most common edible (actually, drinkable) product from that time about

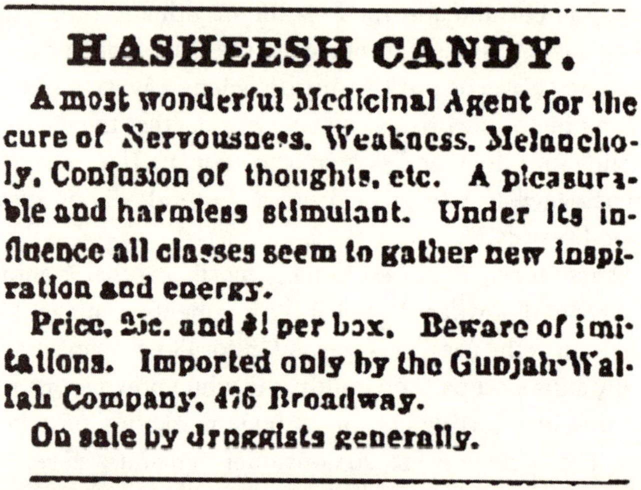

HASHEESH CANDY.

A most wonderful Medicinal Agent for the cure of Nervousness. Weakness. Melancholy. Confusion of thoughts, etc. A pleasurable and harmless stimulant. Under its influence all classes seem to gather new inspiration and energy.

Price, 25c. and $1 per box. Beware of imitations. Imported only by the Gunjah-Wallah Company, 476 Broadway.

On sale by druggists generally.

Figure 7.6. Advertisement for Hasheesh Candy in *Vanity Fair* from 1862.

3,000 years ago is called bhang. Chopra offered a veritable cookbook of bhang recipes, prepared with cannabis leaves and flowers, milk, and a variety of other ingredients such as almonds, pistachios, rose petals, mint leaves, garam masala, ginger, fennel, anise, cardamom, rosewater, and honey.

The more recent history of edible cannabis started 150 years ago and has led to highly creative culinary developments more recently, with clandestine cooking in between. Lacking candy with marijuana cannabinoids has a bit of history. In the 1860s candy lovers were treated to a product offered by the Gunjah Wallah Company in New York called Hasheesh Candy (fig. 7.6). This maple-flavored sugar candy was sold legally for over forty years and was even advertised in the proto-Amazon Sears, Roebuck catalog. It was marketed as fun, sweet, and perfectly harmless. In fact, several other products with psychoactive compounds were offered during this time, including opium- and cocaine-infused tonics that were used medicinally.

The advertisement for Hasheesh Candy (see fig. 7.6) offered it as a medicinal product. In 1906 the United States passed the Pure Food and Drug Act. Although this law was initially ineffective in curbing quack remedies containing psychoactives, it eventually drove most edible marijuana products underground. In 1954 Alice B. Toklas included a recipe for hashish fudge in her famous eponymous cookbook. Needless to say, the publication of this recipe was controversial at the time, and for the most part underground cooking of edible marijuana remained the only way to make it available in this form.

Marijuana-laced food (especially brownies) was produced by users of cannabis during the latter half of the twentieth century as an alternative to smoking. Today cooking with marijuana has become an art and an industry. Even cable channel streaming services have created cannabis cooking reality shows such as *Cooking on High* and *Bong Appetite.* Laced candies are particularly popular, with gummy bears being one of the more popular ways of consuming THC. Hasheesh Candy, in other words, is back in vogue.

8

THC and CBD in the Body

Through decarboxylation, we now have the THC and CBD in either smoke or edible form ready for the body. Where do they go from there? First let's follow a molecule of THC emanating from a hit on a joint, and then do the same for a molecule of THC in a piece of white chocolate. We need to consider how smoke egresses to the body differently than food does for one simple reason: inhaling THC and CBD delivers it to the lungs, while ingesting these compounds in food delivers it to the stomach.

If you take a drag of cannabis from a joint and inhale the smoke, most of it goes into your mouth and down your trachea. While the smoke is in your mouth, there are a large number of volatile chemicals entering it with which your taste buds will interact. Your olfactory system will also detect quite a few odors. This is because your gustatory- and olfactory-sensing systems both work by reacting with chemicals. Your sense of taste is bombarded with a lot of the volatilized terpenes that are wafting into your mouth. The tastes are sometimes overwhelming and are accentuated by your olfactory system, because a good deal of taste is actually smell.

For a comparison, try this simple experiment: Find a cherry- or strawberry-flavored jelly bean. Hold your nose closed, pop the jelly bean into your mouth, and chew. Because the jelly bean is mostly sugar, your taste buds will immediately detect sweetness, but recognizing the taste as cherry or strawberry is not entirely complete until you remove the grip on your nose. Once you do open your nose, you'll immediately get a waft of cherry or strawberry odor that makes you think you are tasting those fruity tastes (as artificial as they might be). Odor assists taste to produce our perception of flavor in our brains.

Taste and smell work through the binding of chemical compounds to tiny molecules embedded in the membranes of some of the cells in your mouth and nose, respectively. The small molecules that make it to your tongue and nose are all different sizes and shapes. Cannabis has not only cannabinoids but also many other molecules such as terpenoids, which are rather small (but in the same size range as THCA and CBDA). The terpene molecules and the cells in your olfactory system act like cards and card readers, respectively, to get the information to your brain about the odors and tastes. Part of the small molecules that do the sensing poke out of the cells to the outside, and some of the molecules poke into the inside of these smell and taste cells. Terpenes (and other molecules—see fig. 8.1) interact with small membrane-bound proteins on the cell surface. These proteins are called receptors, and a chemical signal is produced that is turned into an electric potential; this electrical signal is sent to your brain, where it is interpreted as a particular odor or taste.

Your taste-sensing cells detect five basic kinds of taste—sweet, salty, sour, bitter, and umami—and this means that you have five basic kinds of molecular receptors for taste in your gustatory system. Your olfactory system, on the other hand, has hundreds of these membrane-bound proteins, and because these receptors can act in a combinatory fashion, millions if not billions of different odors can be detected by the typical human olfactory system. The system is so fine-tuned that molecules that look almost identical to one another in three dimensions can be discerned as different, because the angles that the molecules' carbon atoms align with each other are different from terpene to terpene. Figure 8.2 illustrates four terpenes with identical chemical formulas ($C_{10}H_{16}$) and with very similar (but not identical) three-dimensional structures. Although your gustatory system might not detect a difference in taste of these four terpenes (they mostly have citrusy tastes), your olfactory system will detect the slight structural differences of these molecules and interpret them in your brain as different odors (specifically turpentine, herbal, herbal/citrusy, and citrusy). These olfactory signals will also impact the overall sensation of taste.

Taste and smell interact to give us an overall sensation and mental picture for the foods, beverages, and drugs we ingest. Figure 8.3 shows four terpenes that

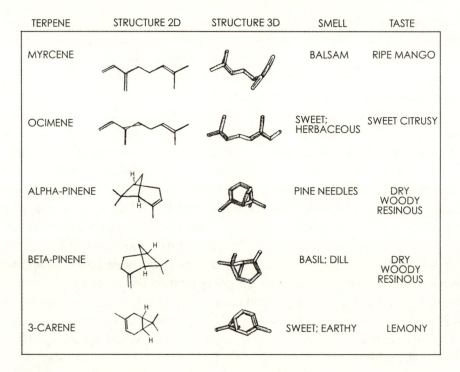

TERPENE	STRUCTURE 2D	STRUCTURE 3D	SMELL	TASTE
MYRCENE			BALSAM	RIPE MANGO
OCIMENE			SWEET; HERBACEOUS	SWEET CITRUSY
ALPHA-PINENE			PINE NEEDLES	DRY WOODY RESINOUS
BETA-PINENE			BASIL; DILL	DRY WOODY RESINOUS
3-CARENE			SWEET; EARTHY	LEMONY

Figure 8.1. Terpenes inhaled in a hit of marijuana. All the terpenes shown here have the chemical formula $C_{10}H_{16}$. Despite having the same chemical formula, they all have substantially different three-dimensional structures. When two chemicals have the same chemical formula but different three-dimensional shapes, they are called isomers.

do not have the same chemical makeup (are not isomers) as the ones shown in figures 8.1 and 8.2. Their chemical formulas are shown in parentheses after the names (nerolidol, linalool, alpha-humulene, and beta-caryophyllene) of each terpene. The overall shapes of these terpenes are notably different, making the receptors they trigger quite different and in turn the tastes they evince quite distinct. When combined with the different odors for these four terpenes, a plethora of combinations of smells and tastes is created.

It is possible that some individuals can taste more or less intensely than others. Perhaps you are one who abhors spicy food or hoppy beer. Or you could be someone who rarely if ever is put off by the taste of the food you are eating. But almost everyone has the same five kinds of taste receptors, so it can't be that someone sensitive to bitter or spicy food or tolerant of all kinds of foods has more

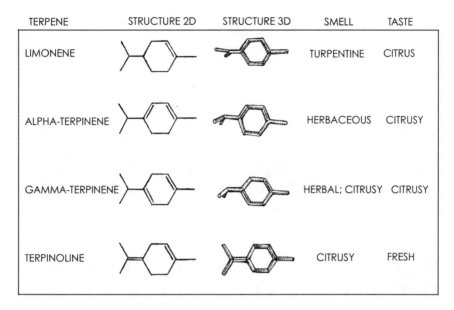

TERPENE	STRUCTURE 2D	STRUCTURE 3D	SMELL	TASTE
LIMONENE			TURPENTINE	CITRUS
ALPHA-TERPINENE			HERBACEOUS	CITRUSY
GAMMA-TERPINENE			HERBAL; CITRUSY	CITRUSY
TERPINOLINE			CITRUSY	FRESH

Figure 8.2. Four terpenes with identical chemical formulas and similar two-dimensional shapes. The central carbon ring is depicted as flat on the surface of the page and shows how the two methyl groups (CH_3) on the left side of the molecule are arranged differently in the four terpenes. The methyl group is twisted into different orientations in the four terpenes. All these terpenes share a chemical formula of $C_{10}H_{16}$ and are isomers of each other, as well as isomers of the molecules in the previous figure.

or fewer of the receptors. Then how do some people end up tasting more intensely or less sensitively than others?

One old notion about how we taste held that different areas of the tongue were responsible for different tastes. For instance, the tip of the tongue was thought to be the area where we tasted sweet, while the back of the tongue toward our gullet is where we supposedly tasted the five categories of the gustatory experience. But our tongues are covered with a field of small mushroom-shaped structures called fungiform papillae—normally about 10,000 of them at birth. Each papilla has cells with all five of the different kinds of receptors embedded in the membranes of each cell. Hence each papilla (also called a taste bud) has the capacity to sense all five major categories of taste. Thus the idea that there is regional control of tastes by the tongue flies out the window. As with our hearing, we lose some of the papillae as we age, and by doing so lose some of our taste.

These small structures are bathed with molecules from the food we ingest,

TERPENE	STRUCTURE 2D	STRUCTURE 3D	SMELL	TASTE
NEROLIDOL (C15H26O)	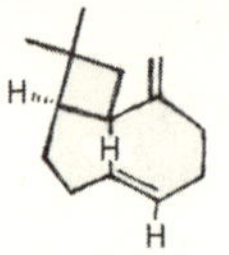	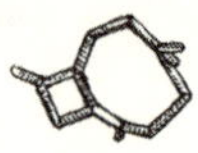	WOODY; APPLE	EARTHY; SWEETNESS
BETA-CARYOPHYLLENE (C15H24)	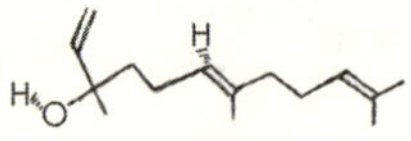	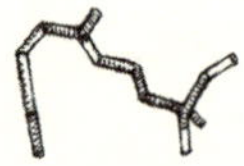	WOODY	SPICY; PEPPERY
ALPHA-HUMULENE (C15H24)	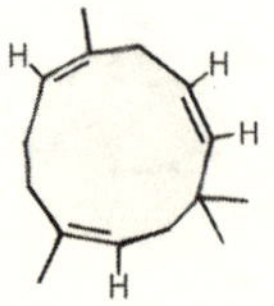	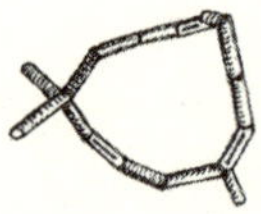	SPICY	HOPPY BEER
LINALOOL (C10H18O)	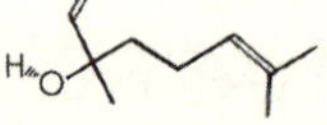	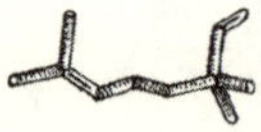	LAVENDER	FRUITY

Figure 8.3. Four terpenes that are not isomers of the previous nine. In other words, they do not have the chemical formula $C_{10}H_{16}$. On the other hand, beta-caryophyllene and alpha-humulene are isomers of each other. These larger terpenes take different three-dimensional shapes.

and these molecules find places to attach to the receptors embedded in the cell membranes of our taste buds. When bound to a molecule such as salt, the receptors in the taste bud cell membranes will produce a cascade of reactions in the cell that soon sends a signal to the brain. It isn't the kinds of receptors that enable some people to taste more intensely than others; it's the number of the papillae that imparts the capacity to taste intensely (or hardly at all). An intense taster is also called a supertaster or hypertaster, and a person who can stand any taste is called a subtaster or hypotaster. A person in between is considered a normal taster. A supertaster will have more than thirty or so papillae in a circular area with a one-centimeter diameter; a subtaster will have fewer than ten; and normal taster will be in between.

A typical population in the United States will include about one-quarter supertasters, one-quarter subtasters, and half normal tasters. Although THC has little to no taste, CBD tastes quite bitter. Supertasting will definitely influence the taste of CBD and make it seem overly bitter. The taste sensations of CBD in normal tasters and subtasters are more than likely not intense, even when the CBD is concentrated into oils. But it always tastes better to mask the taste of a bitter compound (like CBD) with something sweet when ingesting it. If you are a supertaster, you might need more masking of the bitter taste than other people when ingesting it.

On to the Lungs

We may have lost track of our THC molecule with this diversion into taste. But it is still wafting along with all the terpenes previously mentioned and many other molecules. Cannabis is known to make an enormous number of chemical compounds, and will be volatile in inhaled smoke. A weird phenomenon occurs when THC interacts with the cells in your mouth. When THC first hits the mouth, it interacts with the gustatory receptors and then wafts into the nose to interact with the odorant receptors. You will get a bud smell and a floral taste. But THC also will interact with the salivary cells in your mouth and produce the odd-sounding physical state of xerostomia. This state is caused because THC shuts down the production of saliva, causing what is otherwise known as cottonmouth. Saliva is as efficient as any mouthwash in cleansing the mouth of food particles and bacteria that will cause odors and result in bad breath. In addition to causing cottonmouth, marijuana smoke will leave residues of plant toxins, tar, and an array of other stinky molecules. "Weed breath" is thus amazingly unpleasant both for the breather and the person being breathed on.

But let's say our single THC molecule makes it past the receptors in the mouth and doesn't interact with the salivary cells or contribute to xerostomia. What happens next? It enters the upper airway and, oddly, has two impacts that are somewhat opposite of each other. The THC and other molecules in the hit you took can cause coughing, wheezing, and sputum production. These adverse effects occur because THC and other molecules are irritants, and the cells in the upper airway recognize them as such. But once the THC interacts with the cells a little longer, it activates upper airway hyper-responsiveness, anti-inflammatory activity, and antitussive activity (cough suppression). Our THC molecule thus gets past the lining of the upper airway and travels to the lungs.

Our lungs are an architectural wonder. They are made of three lobe-like structures on the left and two on the right. Tubes large and small are distributed

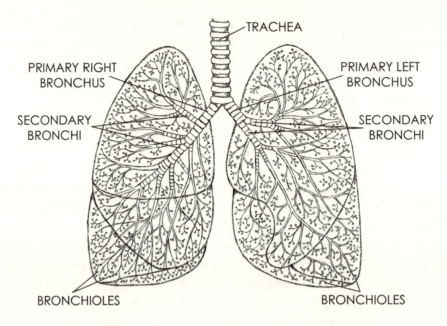

Figure 8.4. Drawing of a healthy lung showing primary and secondary bronchi and bronchioles.

throughout the organ, with the major feeding of the marijuana smoke occurring through a large tubelike structure called a bronchus (one on the left and one on the right). These primary bronchi then branch into hundreds of smaller tubes or secondary bronchi, which then branch into even smaller tubes or bronchioles. At the end of the bronchioles are microscopically small sacs, the alveoli. The alveoli are important in the primary function of the lung, because they are where gas exchange occurs; oxygen is pushed into the bloodstream, and carbon dioxide is pulled out by exhaling. Gas exchange enables oxygen to travel throughout the body in the bloodstream, with some of it going to the brain. Our lucky THC molecule makes it to a small alveolus at the tip of a bronchiole (fig. 8.4), which absorbs it and allows the THC to jump the alveolar barrier (because it is a small molecule) and enter the bloodstream. Once in the bloodstream the THC molecule travels rapidly to other organs. The THC molecule will make it to the brain rather quickly, where it jumps the blood–brain barrier and enters our nervous system. I have dedicated chapter 9 to its further psychoactive journey, but will take the molecule through your stomach here.

The Stomach

We are now onto our second THC molecule, which comes from a food such as ingested white chocolate that has been laced with ground decarboxylated bud. The chocolate tastes sweet, and the sweetness and vanilla taste mask most of the somewhat bitter herbaceous taste that THC will impart to the gustatory and olfactory systems. In the mouth our second THC molecule is confronted with the massive mechanical disruption of the teeth. While maceration occurs, the salivary glands start to produce enzymes. Amylases and lipases are two enzymes in the saliva, but these enzymes will avoid the THC molecule and instead degrade the sugars and long-chain carbohydrates of the white chocolate, making the THC molecules more available later in the process. Past the tongue, the macerated material containing our THC molecule gets to the esophagus. The esophagus is a tubelike organ with two jobs—to get food to the stomach and to keep stomach acids from getting out. It accomplishes the first by muscle movements called peristalsis, and the latter by a sphincter that resides at the junction of the esophagus and stomach.

In the stomach the THC molecule is bombarded with acids and digestive enzymes. But our lucky molecule reaches the bloodstream before these enzymes have an impact on it, by being absorbed by the cells of the stomach. Once absorption occurs, the molecule will pass into the bloodstream easily. Even if it passes through the pyloric valve (the valve that connects the stomach to the small intestine), it will more than likely be absorbed by the small intestine and transferred to the blood, or it will get absorbed in the small intestine and shuttled onto the liver. If it makes it past the small intestine it will end up in the kidneys and eventually in your urine or will be eliminated in your feces. A surprising amount makes it to both of these exits, which is why if you are using cannabis and are tested for drugs, you will most likely lose out to cannabis.

After passing through the stomach, our second THC molecule will probably travel first to the liver. The liver functions as a detoxification device for whatever you have ingested. There our THC molecule might encounter a protein called cytochrome P450 and other cytochrome enzymes called CYP2C9 and CYP3A4, which detoxify most of the molecules such as THC. But let's posit that our lucky THC molecule avoids detoxification by these CYP molecules. If so, other liver enzymes will transform some of the remaining THC into slightly modified forms

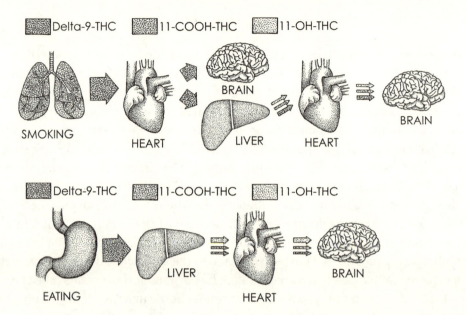

Figure 8.5. The top figure shows the route of metabolism when delta-9-THC is inhaled into the lungs. The thickness of the arrows indicates the relative abundance of that molecule being transmitted from one organ system to the next. The bottom figure shows the route of metabolism for ingested delta-9-THC. The molecule is transported to the stomach, where it is absorbed and sent on to the liver. In the liver there are major chemical transformations of delta-9-THC to 11-OH-THC and 11-COOH-THC. These molecules are then transported to the heart and on to the brain. The pathway through the stomach therefore transmits mostly the highly psychoactive 11-OH-THC and smaller amounts of nonpsychoactive 11-COOH-THC and moderately psychoactive delta-9-THC.

of THC. The THC that gets into our bodies by inhalation or ingestion is called delta-9-THC. The THCs that are metabolized from delta-9-THC in the liver are delta-11-OH-THC and delta-11-COOH-THC (fig. 8.5). The former is more psychoactive than delta-9-THC, and the latter is not psychoactive at all. These molecules are moved to the circulatory system, where they flow about the body much in the same way that our inhaled THC molecule did, reaching many organ systems. Our second THC molecule makes it to the arteries of the brain, where it jumps the blood–brain barrier and enters our nervous system. Its brethren delta-11-OH-THC molecule gets there too simultaneously. For now we will simply state that the THC molecules which make it to the brain will interact with the cannabinoid receptors there.

The dynamics of THC delivered to our bodies by ingestion differ from the dynamics of THC delivered by inhalation. First, with inhalation the THC bypasses the liver and can go directly to the brain. With ingestion, the pathway goes through the liver, and this takes more time, which is why when cannabis is inhaled the psychoactive effects are immediate. It is also why when you ingest cannabis you need to wait for it to take effect. A second dynamic is that the more psychoactively powerful delta-11-OH-THC is in tenfold higher concentrations when ingested relative to inhaling. A final dynamic common to both ingestion and inhaling is that THC can also bind up in fat tissue, where it can lie dormant for several days. Eventually the THC is released slowly back into the bloodstream from these fat deposits. This slow release is why if you are tested for THC in your system a week after your last toke, you might still be positive for cannabis consumption.

So far I have not discussed CBD's journey through the body. But because it is about the same size and shape as THC, what I have described about the movement of THC is generally true for CBD. The only difference is the way that receptors interact with the two molecules in the brain and elsewhere. The most impactful destination of CBD molecules is also not necessarily the brain at all.

What about the Skin?

CBD will enter the body when smoked and when eaten, the same way as does THC. The primary destination for THC is the brain. When smoked, it enters the bloodstream via the lungs, and when eaten by means of the stomach and liver. In the long run, some CBD will make it to the brain the same way THC does, but the cannabinoid receptors are not the only places where it ends up. Instead, it seeks out other kinds of receptors in the brain such as adenosine receptors, serotonin receptors, and vanilloid receptors. Adenosine receptors are involved in the expression of anxiety. Serotonin receptors, when activated, have an antidepressant effect. Vanilloid receptors are involved in inflammation and pain reception. But there is another way to enter the body for which CBD seems to be well suited—topical application. Because some of the medicinal uses of CBD are for localized parts of the body, it makes sense to apply it near the injured or painful site and not by treatment to the entire body. So how does topical application of CBD work? Let's start by considering how organisms distinguish between what is inside them and what is external.

Bacteria, archaea, and single-celled eukaryotes (cells with a nucleus) all have cellular membranes delimiting their insides from the outside. Different cells have different functions, and such functions are best carried out separate from other cells; keeping these functions on the inside of cells does the trick quite well. To accomplish this, cells have membranes and also have evolved ways of regulating what kind of communications can be made between cells. Cell clusters that have developed this way of communicating are usually called tissues—including neural tissue, stomach tissue, and skin tissue. Multicellular organisms such as humans also have a need for keeping the inside from the outside on an organismal level.

Scientists have only recently learned about the evolution of skin. Obviously single-celled organisms lack skin, so researchers need to look for protoskin, or the origin of skin in multicellular organisms. Plants have skin, but it's not the same kind of skin animals have. It is composed of a single cell layer covered with a waxy polymer that separates the internal organs and other parts of the plant from the outside world. Although human skin is different from bird or lizard skin, most mammalian skin is similar to ours, except that we are much less hairy than most mammals. But the cells that make up the skin, the various layers of skin, and the kinds of molecules floating around in the cells of the different layers of skin are similar to those of other mammals.

Skin is our first line of defense in making sure we keep the outside from getting inside (and good things from getting out). In addition to the mechanical exclusion of bad things, like bacteria or toxic molecules, skin is an immunologically active tissue. There are three major layers: epidermis, dermis, and an area that is collectively called the subcutaneous area (fig. 8.6). The epidermis is made up of specialized cells called keratinocytes because they have abundant amounts of a protein called keratin. Keratin is a "jack of all trades" fibrous protein and amazingly versatile. It is used by vertebrates in all kinds of structures like hair, fingernails and toenails, feathers, horns, claws, scales, and hooves. Its role in the cells of these tissues is structural, and in fact the protein provides tensile strength to the cell and hence to the tissue. The epidermal keratinocytes form sublayers, of which the most important for topical application is called the stratum corneum. Imagine a brick wall with mortar that is used to seal the spaces between the bricks. In the stratum corneum the cells (also called corneocytes) are the bricks, and a lipid matrix serves as the mortar to create a rather efficient barrier against the outside world. Just below the stratum corneum is a structure called the basal lamina, which also is fibrous and makes an effective buffer from outside to in. The rest of the cells of the epidermis contain many active proteins and receptors, which

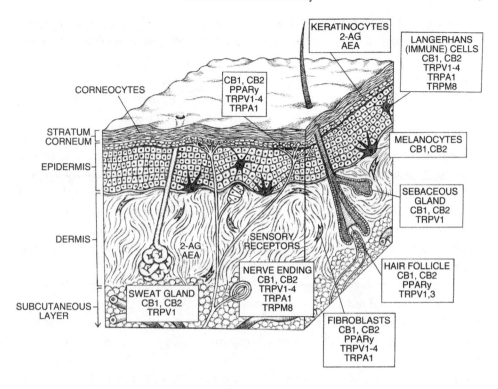

Figure 8.6. Diagram of the skin. The relevant receptor molecules and cell types are shown in the diagram. The subcutaneous layer is below the dermis.

could be the destinations for our CBD molecule. Immune cells are also present in this part of the epidermis.

The dermis comes next, where hair follicles, sweat glands, and nerve endings reside. Our CBD molecule has many potential destinations in this area of the skin too, as all the major receptors for it are floating around this layer. The dermal layer also is home to several kinds of touch sensory cells such as Merkel cells, and Pacinian and Meissner's corpuscles. The subcutaneous layer, the last layer of the skin, contains fatty tissue for insulation and acts like padding when mechanically disturbed. The fat in this layer is also a source for energy.

This skin structure is a lot for our CBD molecule to deal with. But what might be the destiny of our CBD molecule when introduced to the skin via topical application? The molecule hits the skin in the form of a topical oil or spray and is rubbed in. When it hits the stratum corneum, it has two routes it can take to get by that brick-and-mortar barrier. (If it doesn't, it will remain on the surface and more than likely degrade or dry away, prematurely ending its journey.) The two routes are transepidermal (through the cells of the epidermis) and transappendageal (through the interstitial areas between the epidermal cells). One can think of this latter route as finding an area of the epidermis that needs tuckpointing; the CBD molecule will sneak through before the cell can tuckpoint the damage. The transappendageal route simply refers to the microscopic appendages that are embedded in the epidermis and dermis such as follicles, sweat glands, and sebaceous glands. These appendages sometimes aren't perfectly embedded and leave small openings for CBD molecules to get through. Both of these routes pose challenges. But it turns out that oil and water can soften up the skin's layers, so our CBD molecule gets through the stratum corneum and traverses the rest of the epidermis without a problem.

Our CBD molecule is now in the skin's dermis, where hair follicles, sebaceous and sweat glands, and nerve endings reside. These all provide potential breakthrough points with which our CBD molecule can penetrate farther into the tissue it was intended for. There are also plenty of receptors for a CBD molecule to interact with in this region of the skin. How CBD and THC find and interact with these receptors is another set of stories related in chapters 10 and 11.

9

A Plant of (More Than) 1,001 Chemicals

The cannabis plant generates a huge range of chemicals that it synthesizes. Cannabis is so effective at making these chemicals that it is called "the plant of 1,001 chemicals." Actually, a more accurate estimate of the number of chemicals that cannabis produces is greater than 1,600. They are technically called phytochemicals. There is a long history of studying them; many of the molecules that reside in plants are there for good reasons. Two of the most common reasons are for defense (if the plant tastes bad, it won't be eaten) and to attract pollinators and dispersers (if the plant is attractive to animals that move around a lot, it will be fertilized and dispersed). In an evolutionary context, some of the phytochemicals that plants produce are successful at first, but the defensive ones are often in an arms race with the animals that eat them. The race starts with the plant producing a defensive molecule that an insect or larger animal finds noxious or even lethal. Insects will at first stay away from the plant's defensive chemical barrage, but through evolutionary time, mutations in the genomes of the animals that the plant is warding off can occur that will boost resistance to the noxious molecule. The animals can now take advantage of the plant without getting sick or dying.

The cycle will start over again when the plant escalates the arms race by evolving a more efficient version of the first molecule or even a brand-new noxious chemical. The vast number of plant secondary compounds that exist as defensive chemicals is good evidence of the lack of truces or even détentes in this arms race. In many cases a race is evident from understanding how animals develop resistance. Because a lot of these evolutionary cycles involve insulting the taste buds of other organisms, many plant phytochemicals are bitter, oily, tough to digest, or downright toxic to the animals eating the plant. And many of these defense and attraction phytochemicals have been used by humans for a wide range of other purposes.

Of the 1,001 chemicals, the major categories that are involved in marijuana potency, taste, smell, texture, and overall pleasantness are flavonoids, lignins, sesquiterpenes, triterpenes, monoterpenes, and cannabinoids. Cannabinoids are of course where the action is when it comes to medicinal and recreational purposes, but the other phytochemicals may turn out to be equally important.

Flavonoids

Each of the six types of chemicals I consider have distinctively shaped molecules that are important to their function. They have a basic skeleton consisting of atoms from the six basic building blocks of animal life: CHNOPS, or carbon, hydrogen, nitrogen, oxygen, phosphate, and sulfur. Often the carbon atoms form many-sided rings that are drawn as hexagon-like figures in chemical diagrams. These hook together to flesh out the basic skeleton of a group of related chemicals such as the flavonoids. Flavonoids have three of these rings in their basic skeletal structure. A flavonoid looks like a small puppy: neck, ears, and head, with two of the rings making up its body and one its head. Luteolin (fig. 9.1) offers a good example; its ears and tail are clearly positioned.

What makes one flavonoid different from the next is based on differences inherent in each "puppy." If the puppy has two ears (those hydroxyl or OH structures coming off the head of the skeleton in luteolin in figure 9.1), then it can be any one of several flavonoids. Puppies with two ears can either be lying down or standing up. Standing puppies can have their heads up or can be "eating" with their heads down. The upright puppy can either be standing still or jumping. All of these different puppies occur when there are very slight changes to the basic

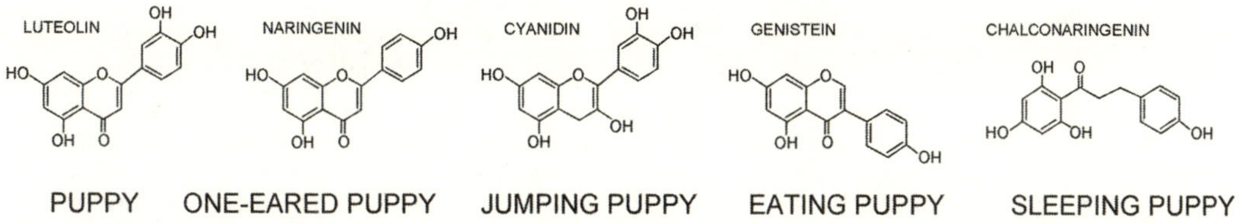

LUTEOLIN   NARINGENIN   CYANIDIN   GENISTEIN   CHALCONARINGENIN

PUPPY   ONE-EARED PUPPY   JUMPING PUPPY   EATING PUPPY   SLEEPING PUPPY

Figure 9.1. Flavonoid "puppies." Morphing from one puppy to the next can be accomplished with simple chemical changes. For instance, the basic puppy can change into a one-eared puppy by losing one of the OH groups on its head. The "sleeping puppy" is shown on its back.

skeleton, such as adding or subtracting an OH group. In all, there are over 5,000 puppies that have been characterized as flavonoids from nature.

Just as there are breeds of dogs, there are also major basic categories of flavonoids. There are five major ones: flavones, flavonols, isoflavones, anthocyanins, and chalcones. Each is characterized by having a distinctive basic skeleton that has slight alterations to it, as described previously for the puppies.

Flavonoids are found in a broad array of plants—both fruits and vegetables. They are thought to be associated with broad-spectrum health-promoting effects and are included in a wide range of medicinal products. The wide-ranging medicinal characteristics of flavonoids reside in their chemical nature and shapes— those ringlike structures that make up the puppies' bodies. Those ring structures (called polyphenolic structures) render the molecules antioxidative, anti-inflammatory, and anticarcinogenic. Flavonoids are also important players in two of the plant kingdom's most showy characteristics—aroma and flower color. In cannabis the flavonoids are found in all parts of the plant and are probably an important aspect of its overall aroma and taste.

Terpenes

Terpenes are another group of natural plant products. Perhaps the most famous terpenes are natural rubber and paclitaxel (Taxol), an anticancer agent. All terpenes have a similar chemical equation with five carbons and eight hydrogens repeated n times over and over [$(C_5H_8)n$, where n can be any integer and refers to the number of times the C_5H_8 can be repeated]. When n is two, making a molecule that is $C_{10}H_{16}$, the terpene is called a monoterpene. When n is four, making a

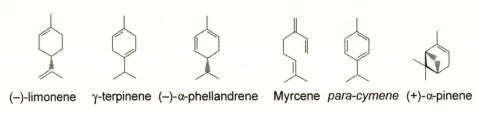

(–)-limonene   γ-terpinene   (–)-α-phellandrene   Myrcene   *para-cymene*   (+)-α-pinene

Figure 9.2. "Bar dart" monoterpenes.

molecule that is $C_{20}H_{32}$, the terpene is called a diterpene. When n is six it makes a molecule that is $C_{30}H_{48}$, which is called a triterpene. Most terpenes have even-numbered n's, except for when n is three and makes a molecule $C_{15}H_{24}$ called a sesquiterpene (sesqui = 1.5). The number and positions of the carbons and hydrogens dictates the characteristics of the terpene, with over 30,000 of these kinds of molecules found in nature.

Monoterpenes mostly resemble bar darts with fat barrels; some that are relevant to our understanding of cannabis are illustrated in figure 9.2. Even those that don't look like darts (such as pinenes) still have a dart-like tip. Myrcene, limonene, and pinene are all monoterpenes that influence the taste and aroma of cannabis. They are also important because the way they are generated in plant tissues is through two biosynthetic pathways: the MEP (methylerythritol-4-phosphate, a precursor molecule for the pathway) and the MVA (mevalonate, another precursor molecule for this second pathway) pathways. Some of the proteins that the plant genome makes are important enzymes in these biosynthetic pathways, which are also significant for the synthesis of the cannabinoids. One category of enzymes for which plant genomes code are the monoterpene synthases. These take the preliminary chemical structures and transform them to the terpenes, which give the plant its texture and aromatic and gustatory characteristics. The monoterpene synthases result in important compounds such as geraniol, myrcene, and limonene.

The triterpenes ($C_{30}H_{48}$) look a lot like five-segmented centipedes. These molecules have a basic segmented chemical skeleton with side groups on one end that resemble antennae (fig. 9.3). As with the flavonoids, modifying the basic structure of these centipede skeletons with slight changes will yield different kinds of triterpenes.

Why are we so interested in the shape of these chemicals? The shape of a molecule has everything to do with its odor, taste, and other recreational or me-

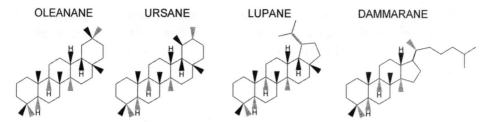

Figure 9.3. Centipede-like triterpenes. Four basic kinds of triterpenes are pictured, with their basic skeletons illustrating the chemical equation $C_{30}H_{48}$.

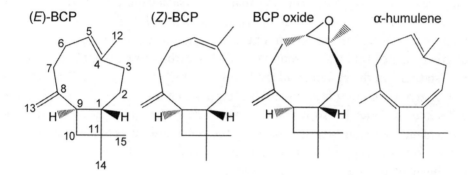

Figure 9.4. Some stop-sign sesquiterpenes inhaled in a hit of marijuana. All the terpenes here share the chemical formula $C_{15}H_{24}$. Despite having the same chemical formula, they all have different three-dimensional structures.

dicinal properties. So subtle differences in some of these molecules will often change the landscape of how these chemicals influence our bodies. For instance, all monoterpenes have the same chemical formula—$C_{10}H_{16}$; they are isomers of each other, which means they all have the same number of carbons and hydrogens. If you examine the structure of these terpenes closely, although you will see some similarities in their three-dimensional structure, you will also observe that they differ from one another. It is these differences that enable our taste and olfactory senses to tell many of them apart.

Other terpenes cannabis produces that have drawn attention are beta-caryophyllene (*E*-BCP) and humulene (fig. 9.4). No doubt about it, that skeleton is a stop sign. These kinds of terpenes are sesquiterpenes with the chemical formula $C_{15}H_{24}$. *E*-BCP has recently been reported to bind to cannabinoid receptors, suggesting that it is not only a terpene but also could be considered a cannabinoid. *E*-BCP is recognized by our taste buds as peppery. But alpha humulene triggers a hoppy taste. The slight difference in the structures of the two molecules makes all the difference to our taste and olfactory receptors.

Lignins

This category of major plant secondary compounds is the most complex. There is no single lignin skeleton for the basic structure of this category of molecules, and they are cobbled-together structures that are physically quite rigid. Their rigidity renders them suited for structural purposes in plants. Lignins resemble subway maps, with the various side chains of the different molecules being the red, blue, or green lines of a subway system (fig. 9.5). The carbon rings in lignin function as the "stations" in the system, and the bonds are the "tracks." Just as Tokyo's subway system is different from the Paris Métro, which is equally different from the MTA in New York City, so are the lignin molecules that give structure and rigidity to plant tissues, mostly by shoring up the cell membranes of plant tissues. Although the primary utility of lignin in plants is structural, these molecules can also be involved in texture, odors, and taste.

The beauty of these phytochemicals lies not in their simplicity of structure (they can be quite complex structurally, as the lignins demonstrate), but rather in the simplicity with which they are generated and the simplicity with which a large degree of novel molecules can be created with only slight changes in side chains and specific parts of the chemical skeleton.

Cannabinoids

There are two major synthetic endpoints for this category of secondary compounds: THCA and CBDA, which are the stars of this book. These two chemicals were discussed in some detail in chapter 7 when we addressed decarboxylation—that magical chemical step that converts THC and CBD acid precursors into their decarboxylated potent forms. But there I black-boxed most of the molecule (see fig. 7.1) and exposed only the parts of the THCA and CBDA molecules that are the targets of decarboxylation. Let's take a closer look at what is behind those black boxes.

Other plants like those in the genus *Echinacea* also produce cannabinoids,

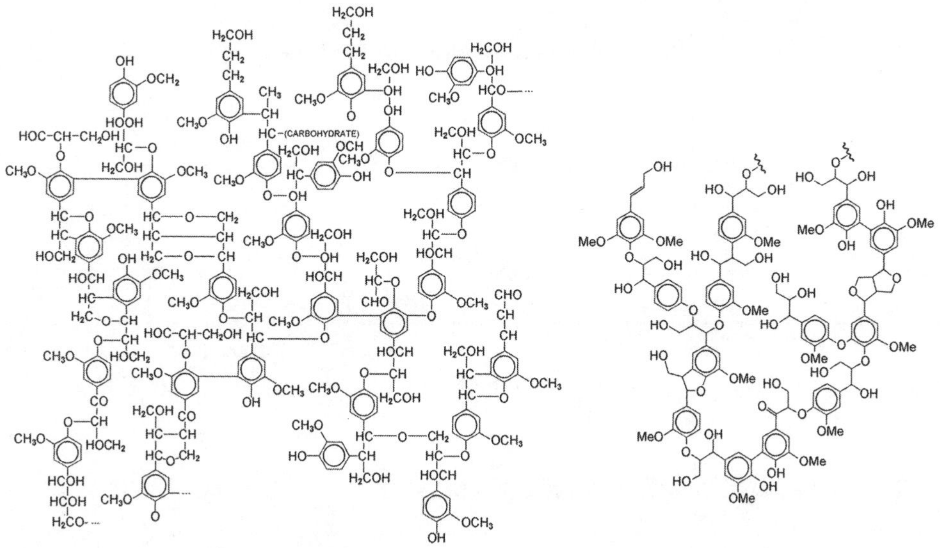

Figure 9.5. Lignin "subway maps." Both of these long-chain lignins are single molecules that have linked many subunits together to form the overall final molecular conformation.

but it is cannabis where the real deals (THCA and CBDA) are found. There are over 100 cannabinoids that have been isolated so far from cannabis. They mostly share the chemical equation $C_{21}H_{30}O_2$. There are ten major cannabinoids in marijuana (fig. 9.6). When I examine the cannabinoid structures that these molecules take, I see the baby alien that popped out of John Hurt's torso in the first *Alien* movie. The creature is complete with a tail, legs for scurrying around, and a bizarre head only a mother could love. The two major cannabinoid endpoints (CBD and THC) are accompanied by a wide range of other cannabinoids that are only now being examined for their impact when ingested. Table 9.1 lists some of the various cannabinoid type classes that have been discovered so far. Of the 480 or so compounds already characterized, 180 can be categorized as cannabinoids, of which 120 are listed in table 9.1. Of these, about 150 are also classified as terpenes.

Recycling

All the phytochemicals discussed so far are produced by the cannabis plant itself. The important starting molecules for phytochemical production in cannabis are

CBGA (CANNABIGEROLIC ACID)

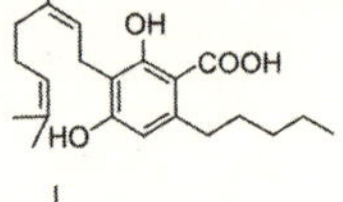

THCA (Δ9-TETRAHYDROCANNABINOLIC ACID)

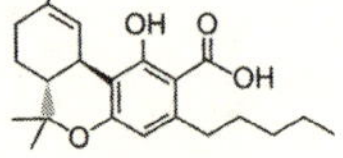

CBDA (CANNABIDIOLIC ACID)

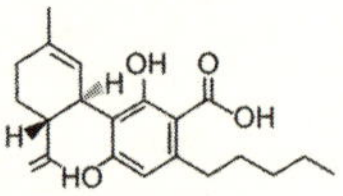

CBGVA (CANNABIGEROVARINIC ACID)

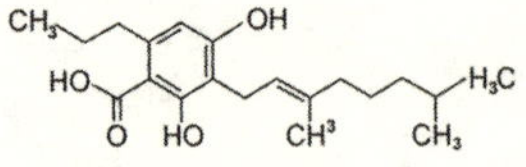

THCVA (TETRAHYDROCANNABIVARINIC ACID)

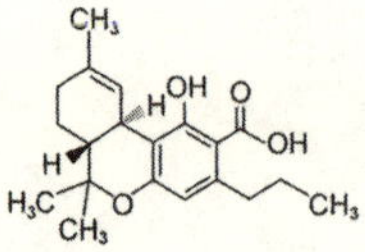

CBDVA (CANNABIDIVARINIC ACID)

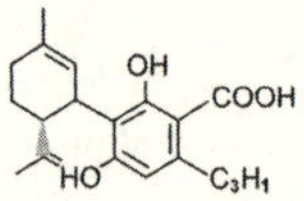

CBCVA (CANNABICHROMEVARINIC ACID)

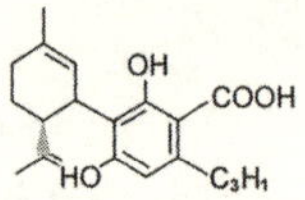

CBCA (CANNABICHROMENENIC ACID)

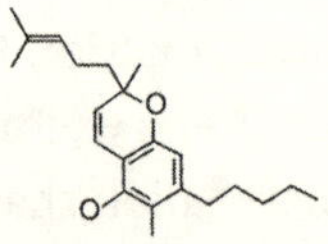

CBD (CANNABIDIOL)

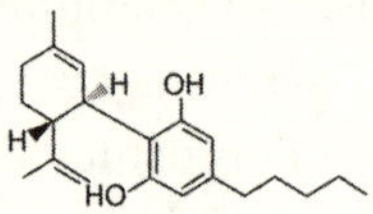

THC (DELTA-9 TETRAHYDROCANNABINOL)

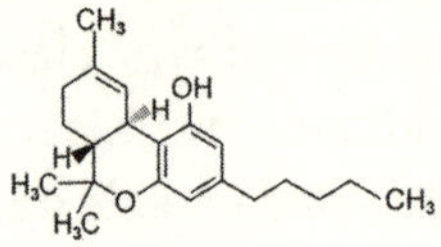

Figure 9.6. Cannabinoid "aliens" with tails to the right and heads to the left. Focus on the two bottom chemicals, THC and CBD. These are decarboxylated forms of the two important chemicals in cannabis. Note that some of the aliens above these two (THCVA, CBDVA, and CBCVA) don't have tails because their active forms don't need them. Also note the longer tail on the CBGVA alien. A precursor, it gives rise to THCVA, CBDVA, and CBCVA.

TABLE 9.1. CANNABINOID TYPE CLASSES, NUMBER OF COMPOUNDS IN THE CLASS, AND EFFECTS

Cannabinoid type class	Number of compounds	Abbreviation	Effect
Cannabigerol	16	CBGA, CBG	Antibiotic, analgesic
Cannabichromene	9	CBC	Anti-inflammatory
Cannabidiol	7	CBDA, CBD	Anxiolythic, antipsychotic
Delta-9-tetrahydrocannabinol	23	THCA, THCCV, THC	Euphoriant
Delta-8-tetrahydrocannabinol	5	Tetrahydrocannabinol	Weak euphoriant
Cannabinol	11	CBN	Sedative
Cannabinodiol	2	CBND	Psychoactive
Cannabicyclol	3	CBL	Nonpsychoactive
Cannabielsoin	5	CBE	Nonpsychoactive
Cannabitriol	9	CBT	Anti–breast cancer
Miscellaneous	30		Several

acetyl CoA, pyruvate, fatty acids (such as hexanoic acid), and small amino acids such as phenylalanine. These small starting molecules are juggled by an array of enzymes to produce larger molecules such as flavonoids, lignins, cannabinoids, and terpenes. The ancestral cannabis plant evolved a preference for certain molecules that turned out to be good starting points for synthesizing the multitude of other chemicals with broader functions. The set of starting materials—the enzymes that transform those raw materials and the steps required—are called pathways. The pathways to flavonoids, terpenes, lignans, and cannabinoids are intricately intertwined with each other, producing a spaghetti-like picture of phytochemical production. Since cannabis relies on all these phytochemicals for its characteristics, the picture of cannabis secondary compound synthesis is particularly convoluted and sticky.

Cells are the ultimate recyclers. They make new molecules that they need for survival by anabolic means (building up) and process old ones that provide the raw materials for larger molecules and produce energy through catabolism (breakdown). The Krebs (citric acid) cycle offers a clear example of how a cell recycles molecules through anabolism and catabolism over and over, using acetyl CoA as the pivotal molecule. Acetyl CoA is important for both building up and breaking down molecules in a variety of vital functions. Its anabolic side connects small molecules together to make the overall bigger molecule of acetyl CoA. Its catabolic side involves the breaking down of acetyl CoA into smaller molecules that are used in generating energy via the Krebs cycle. In addition, acetyl CoA molecules are catabolized into smaller molecules in the synthesis of sesquiterpenes and triterpenes.

There are four major enzymatic pathways that are important in the production of phytochemicals in cannabis (fig. 9.7). We have already discussed acetyl CoA as a starting point for one of these pathways. This pathway occurs in the cytosol of the plant cell and is called the MVA (mevalonate) pathway. The pathway results in triterpenes and sesquiterpenes. An important enzyme in this pathway is farnesyl pyrophosphate synthase, which produces farnesyl pyrophosphate or FPP, an important precursor to both sesquiterpenes and triterpenes. This pathway can also produce malonyl CoA, another coenzyme that is diverse in its function. It can be shuttled to other pathways and be a precursor molecule for cannabinoids and flavonoids.

The second major pathway is called the MEP (methylerythritol 4-phosphate) pathway and is specific to mitochondria. The source molecule for this pathway is pyruvate, a small molecule that results from the cleavage of sugar molecules. The MEP pathway results in monoterpenes such as myrcene and limonene. Two intermediate molecules in this pathway—DMAPP and IPP—are also intermediates in the MVA pathway. Whereas DMAPP is converted to farnesyl pyrophosphate in the MVA pathway and eventually into sesqui- and triterpenes, in the MEP pathway DMAPP is converted to GPP and then to monoterpenes with specific enzymes for the particular monoterpenes.

The third pathway results in the synthesis of flavonoids and lignins and is called the phenylpropanoid pathway. The precursor molecule for this pathway is the amino acid phenylalanine. In a series of complex reactions, this small precursor molecule can be converted to flavonoids. In addition, a side reaction of flavonoid precursors (specifically coniferyl CoA) can be transformed into lignans. Malonyl CoA can also enter this pathway and be a downstream precursor for flavonoids. If phenylalanine does not enter the phenylpropanoid pathway and instead is processed into tyramine, it can be used as a precursor for lignans.

The last but certainly not least of the four pathways is the polyketide path-

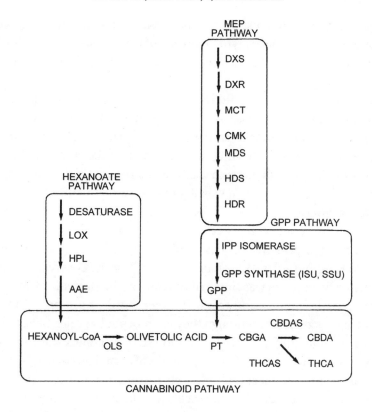

Figure 9.7. The major phytochrome or secondary compound pathways in cannabis.

way. Its precursor molecules can be fatty acids (those long-chain lipidlike molecules), or malonyl CoA can be shuttled into this pathway as a precursor. An important intermediate in this pathway is olivetolic acid, which after a simple reaction produces cannabigerolic acid (CBGA). CBGA is somewhat versatile as it can be converted to cannabidiol (CBDA), cannabichromene (CBCA), and delta-9-tetrahydrocannabinoid (THCA). The ultimate production of CBDA and THCA is accomplished by CBD synthase and THC synthase, respectively. The real beauty of these pathways is that they can sequester their products into different parts of the plant cell, and they seem to be able to feed each other raw materials. What this means for cannabis growers and breeders is that getting a particular strain to produce more THCA or CBDA is not a simple task of overexpressing the synthase genes for these compounds, but instead involves complex interactions that are not so simple. It is remarkable that four major pathways can account for most of the phytochemicals of cannabis. All the major kinds of phytochemicals—lignans, flavonoids, terpenes, and cannabinoids—spring from these four pathways.

In addition, it is not too wild to think that the lignins and especially flavonoids are involved in synergistic relationships with cannabinoids that increase the potency of different cannabis strains. Getting a plant to produce a lot of THCA may not be as simple as controlling the expression of the synthase that converts CBGA into THCA. It could happen that way if a robust synthase arose by mutation that was more efficient than existing ones. But that would require a change in the actual synthase.

Evolution of function has been studied by many researchers. One of the earliest and most informative studies in this realm was undertaken by geneticists Mary Claire King and Allan Wilson in the 1970s. They discovered that the amino acid sequence divergence was minimal in many of the proteins of very different organisms such as humans and frogs. In other words, both organisms had the same protein structures, even though they are morphologically, behaviorally, and physiologically quite different. Their solution to this paradox was to suggest that it's not necessarily the structure of the proteins of organisms that accounts for the big differences in anatomical and other characteristics, but rather how expression of the genes is controlled. In the case of cannabis, while CBDA synthase and THCA synthase are good starting points for increasing the production of these two chemicals, making more efficient synthases may not be the only or best answer.

There is no doubt that the potency, taste, medicinal impact, and other traits of cannabis strains are in some ways controlled by genes. One way to show this is to demonstrate a correlation of DNA sequence variation around these genes with cannabinoid content. That is, to demonstrate that certain DNA sequence variants near the protein coding part of the gene correlate with cannabinoid concentration. Plant breeding researcher Sophie Watts and colleagues were unable to show this correlation with cannabinoid synthase genes. However, they did observe that genetic variation in terpene genes was correlated with cannabinoid labeling (that is, whether a plant was colloquially called Indica or Sativa).

A further reason for expecting a more complicated architecture for the genetic control of cannabinoid content is the "entourage effect." This idea suggests that cannabinoids, terpenes, and other molecules specific to cannabis act in concert with each other to produce the overall medicinal or recreational effect of a strain. The Watts study supports this entourage effect, but some researchers such as pharmacologist David Finlay and colleagues are not convinced of the entourage effect. I will return to this problem in chapters 12 and 13, when I discuss attempts to map the genes involved in cannabinoid concentration using genome techniques, and the impact an entourage effect might have on the medicinal properties of cannabis. I can state here that genetic mapping sometimes points to the THCA and CBDA synthase genes and sometimes elsewhere (usually terpene synthesis genes).

Where Is the Action?

Previously I mentioned trichomes as being an important part of the marijuana plant (see chapter 6). Trichomes develop all over the plant's aerial parts (above soil level), specifically on flower heads and leaves. Trichomes are not specific to cannabis, and they function as little factories and eventually warehouses of plant secondary compounds in a broad array of plants. Certain plants have evolved to have these secretory organs for protection from predators.

In addition to discouraging herbivores, trichomes protect plants from fungal infections, bacterial pathogens, ultraviolet radiation, and high temperatures. There is also a need to sequester some of these phytochemicals away from the rest of the plant because these secondary compounds can be harmful to the plant. Trichomes offer a good solution to this problem of toxicity. Some plants have multiple kinds of trichomes that often bear little resemblance to each other. For instance, trichomes can be either unicellular or multicellular, even in the same plant. Charles Darwin was one of the first biologists to recognize that some organisms would develop or evolve certain structures that looked like structures in other organisms, but that the structures bore a connection by analogy rather than by common ancestry. Darwin called this "convergence" and defined it as similar characters without common ancestry. Thus we should be careful when we state that one trait of an organism is the "same" as a similar trait in another organism. "Similar" and "same" are not interchangeable. Because the definition of trichome is broad and encompasses a wide range of morphologies, this makes convergence of these structures across the plant kingdom a fact of life.

Another interesting case of convergence in animals is eyes. As light-sensing organs, they are grouped together by some biologists and called "eyes" based on their light-sensing capacity. But the eyes of an octopus and the eyes of a mammal are not the same things, despite some structural similarities and their overall func-

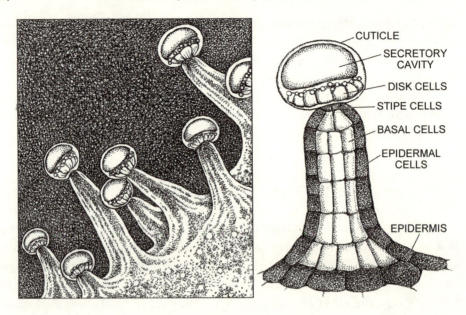

CUTICLE

SECRETORY
CAVITY

DISK CELLS

STIPE CELLS

BASAL CELLS

EPIDERMAL
CELLS

EPIDERMIS

Figure 9.8. Trichomes and trichome structure. From Tanney et al. (2021).

tional similarity (detecting and focusing light). According to some biologists, animal eyes have evolved over twenty or so times. This means that there are over twenty different kinds of eyes in animals. The odd thing, though, is that an octopus eye uses some of the same genes to control its development as a mammal eye. Like eyes for animals, trichomes have undoubtedly evolved several times in the plant tree of life. They are a good example of convergent functional evolution. They have also taken up an important role in the physiology of cannabis plants.

Trichomes develop from the epidermis or surface of differentiating tissues in plants. In *Arabidopsis* the development of trichomes is controlled by a gene regulatory network (GRN), and there is no reason why what applies to *Arabidopsis* trichomes is also the case for cannabis trichomes. How these GRNs change from plant species to plant species determines the timing of development of the structures and what their function will be. Tiny tweaks (like overexpression of one gene in the network) sometimes can change the timing of development and morphology of a plant's trichomes.

Figure 9.8 shows cannabis trichomes protruding from the calyx of a flower. These trichomes develop a stalk and a bulbus head. The head is where the action is, because the disk cells lying right below the secretory cavity are where the monoterpenes, sesquiterpenes, and cannabinoids are synthesized. After synthesis these phytochemicals are transported and stored in the secretory cavity. As noted previously, the secretory cavity serves a twofold purpose: first to store the synthesized phytochemicals it will use for defense, and second to sequester the compounds from the rest of the plant (because the compounds can be toxic).

Cell biologist Samuel Livingston and colleagues used microscope techniques to examine where and when these secondary compounds are produced in the trichomes. They suggest that there is a cluster of cells that actively produce compounds. Livingston and colleagues call this cluster a "supercell." The cells in this supercell are intricately connected to each other and to the rest of the trichome. Its general location abuts the secretory cavity (see fig. 9.8). Livingston's research clearly shows that cannabinoid synthesis is carried out in this area of the trichome and secreted into the cavity.

The trichome is an economic structure; it both stores and synthesizes and leaves little waste to boot. While the disk cells do most of the synthesis, the non-trichome cells that are connected to the flower supply the trichomes with sugars for synthesizing the secondary compounds. The stalk itself acts to both store sugars and to supply the bulbous head with sugars from the floral tissue. Hence there is a warehouse of sugar in the stalk, abutting a factory making secondary compounds in the disk cells, with another warehouse storing the secondary compounds in the secretory cavity abutting that, all within the confines of a trichome. Trichomes are thus where the psychoactive action is, and trichome function holds the key to terpenoid and cannabinoid synthesis in cannabis.

10

Messy Brains and Marijuana

Even when not under the influence of marijuana, our brains are messed up. Neuroscientist Gary Marcus wrote a book called *Kluge* about how messy our brains are. He used the German word *kluge,* because as it is defined ("a clumsy or inelegant solution to a problem") he believed it served as a fitting description of our brains. Although a centralized nervous system anchored by a brain was a major innovation in evolutionary history, the clumsiness of our brains is a part of the story of how marijuana impacts our nervous system. The messiness makes it possible for a lot of places where exogenous chemicals such as THC can step in and "clog" the nervous system. That is one of the hallmarks of a kluge; its function can easily be altered because it is made of many moving parts.

A Planet Overflowing with Brains

The vast majority of organisms on this planet do not have brains; single-celled organisms such as bacteria and protists that make up the bulk of organisms lack them. One of the hallmarks of a nervous system is a synapse. These are points of contact or junctions between cells, across which signals are passed via the movement of chemicals and electrical impulses. Most animals have brains, but sponges (which are also animals) don't have brains or synapses. Placozoa, the tiny pancake animals we met in chapter 3, also lack brains or nervous systems. But both sponges and placozoa have many of the genes in their genomes for making nerve cells, and they implement communication among cells using electrical and chemical means. The kinds of cells in placozoa and sponges that are most likely the ancestors of nerve cells are secretory cells that, upon stimulation of some sort, secrete a protein product. Synapses are found in three of the basal metazoan lineages discussed previously: ctenophores, cnidaria, and bilaterians (see chapter 3). If these three groups have a single common ancestor that is unique to them, then it can be argued that nerve cells evolved once. But there is some controversy about the relationships of these three groups. More important are the structural differences among the synapses of organisms within the three groups. Some researchers have therefore suggested that synapses evolved independently within these groups.

The common ancestors of these groups had a common genetic toolbox for constructing structures important in their biology. Like any tools in a toolbox, they have been used in different combinations to make a wide range of constructions. Any structure built with tools from the toolbox depends on the carpenter and the sequence with which the tools are used. But when a nail needs to be pounded into a board, the carpenter almost always uses a certain kind of hammer. Similar outcomes will use the same tools, but they are independently conceived.

This common toolbox resulted in two outcomes. First, it enabled problems such as cell-to-cell communication to be solved by very specific processes and molecules. It also ensured that when different lineages found a solution to a common problem, similar tools were used by the different lineages to solve the problem. The use of common tools, or genes in this case, does not mean that the organisms solved the problem in exactly the same way, though. Nor does it mean that the solution was passed down from the proto-carpenter. Hence sometimes different lineages have produced different solutions (but using the same tools) that resulted in similar-looking structures. The three lineages—ctenophores, cnidarians, and bilaterians—all diverged from each other at least more than 550 million years ago. The evolutionary solution to nerve cells in cnidaria and ctenophores resulted in a similar overall structure called a neural net, a series of nerve cells connected by synapses that run through the bodies of these animals, but no brain. While both groups can be said to have neural nets, the net in cnidarians is structurally different from the net in ctenophores. Bilateria found yet another solution, which included centralizing the nerve cell communication in a localized structure: a brain.

If ctenophores have synapses and bilateria have brains, what about plants? Do plants have brains? My colleague Ian Tattersall and I have addressed this question several times in our books on wine and beer, two beverages dependent on plants as an ingredient. There are observations that plants can communicate information systemically. They are also thought to communicate with each other and with members of other species and to have "memories" and a capacity to "learn." It is well known that some plants are heliotropic (a specific kind of phototropism)— that is, they respond by altering their motions to optimize intake of sunshine. There are stunning time-lapse videos of sunflowers bending to face the sun and even moving to face the sun directly as the sun moves across the sky. An even more amazing thing happens when the sun sets: the flower returns to its initial position of where it was when the sun first rose. There are many other kinds of tropisms that plants use to navigate their external world. Gravitropism, hydrotropism, thigmotropism, chemotropism, galvanotropism, and traumatropism are all involuntary responses that enable plants to deal with their surroundings. They help plants respond to gravity, water, touch, certain chemicals, electricity, and wounding, respectively. Is tropism "memory"? Perhaps. But plants do not make these responses with a nervous system or a brain. They do it using some other kind of organ or anatomical system.

One way a plant can defend itself is to take advantage of some aspect of its attacker's nervous system. Cannabis, for example, produces cannabinoids and terpenes that interfere with our internal endocannabinoid system. If plants had nervous systems, then it is more than likely they wouldn't be able to evolve neurological solutions to animals that eat them, because those solutions might also be detrimental to their own well-being. Having a nervous system would close them off from a wide range of possible defense mechanisms based on disrupting animals' nervous systems.

One example is neuroreceptors, which act like little keys that unlock proteins on one of the cell surfaces of the neural synapse. The genomes of many plants have genes that code for small molecules called glutamate receptors. These glutamate receptors are used by plants in root and pollen tube development, because they mimic glutamate receptor molecules that animals need for proper functioning of their neural synapses. Glutamate receptors can also poison the nervous systems of predatory animals.

It is not surprising that plants have evolved a completely different way to "pretend" they have nervous systems. After all, animals solved the problem of flight at least four different times: for insects, birds, mammals, and pterosaurs. Each time the animals evolved a common solution (wings), but the common solutions have little to do with each other structurally.

Does having a brain or sometimes multiple brains make an organism "smart"? You may have heard stories about organisms with more than one brain—for example, some leeches have as many as thirty-two brains. (Some leech species are also supposed to have ten stomachs, nine pairs of testicles, and hundreds of teeth!) What is clear is that some leeches have a head brain and a tail brain. To conclude that having multiple brains makes an organism smart might be misleading. After all, would you think it smart to live in the anal passage of a hippopotamus? *Placobdelloides jaegerskioeldi,* the critter that does this, is "smart" because it takes advantage of the blood-rich tissue in the anal passage of these large mammals. But such smartness is simply a reasonable response to natural selection. Just because plants don't have brains doesn't mean they aren't smart in some ways.

Indeed, it is debatable whether expansions or blobbings of the nervous system in certain places are really brains. It might be more appropriate to state that certain regions of a nervous system may appear larger than other usual parts of the neural cord. Humans like to find analogies to help them think about nature. The clump of nerve cells in the tail of a leech serves the purpose of connection and organizing nerve impulses in the leech body, so we think it resembles our brains. But analogy is not a good principle for understanding evolution. Determining what is the same "thing" from organism to organism is a problem that has been recognized by many an evolutionary biologist, not the least of whom was Charles Darwin.

Convergent traits don't necessarily have to be the "same" thing; they just need to look alike or seem to serve a similar function. In fact, it is a good bet that if two organisms have a convergent trait, these really aren't the same. For instance, let's consider trichomes again. These structures appear convergently in lots of plants. As counterintuitive as it might seem, in many cases the trichomes of one group of plant are not the same as the trichomes of another. What is missing is the derivation of those trichomes from a single and most recent common ancestor. If intervening groups of plants (either more closely related to one group or the other) do not have trichomes, then the trichomes of the two groups are not the same. Darwin recognized this as the difference between analogy (looking alike but not the same) and homology (looking alike because of common ancestry)—

one of the most important concepts in evolutionary biology. When scientists say "a leech has thirty-two brains," they usually don't mean brains, but rather structures that centralize neural behavior. While this sounds a little like a brain, it is what its description says it is: a clump of neural cells that human minds may feel more comfortable calling a "brain."

How Did Our Brains Get So Messy?

There is a one-word answer to this question: evolution. You might be thinking that if the human brain evolved, it should be close to perfect. Although natural selection does a good job of finding solutions to problems that organisms must surmount to survive, natural selection doesn't strive for perfection. Instead, natural selection often molds structures and processes in inelegant ways. There are several evolutionary principles that will thwart the perfection of an organism or even one of its traits. The most important is that evolution can only work on already existing variation in populations. Natural selection doesn't simply conjure up the right variation that would lead to perfection. One can get close to perfection for specific traits when doing selective breeding or artificial selection experiments, but nature on its own doesn't work that way. Another factor relevant to neural messiness is that not all evolution proceeds by natural selection. Small population sizes will increase randomness in populations as they change. These random events essentially also thwart perfection, because they will pass on less-fit genotypes, and these will survive as part of the random process of sampling error. Still another good reason perfection doesn't arise is that in the short term, organisms are also stuck with their basic body plans. The most efficient and rapid way to cope with natural selection is to build on that existing body plan. During the development of better and better computers, while a lot of the inner workings were carried over from one generation of computer to the next, engineers who designed them could actually completely scrap an old design and start anew. But again, nature doesn't work that way; it has to shape what already exists.

So humans are left with a brain that has many of the hallmarks of the evolution of vertebrate brains over the past 300 to 400 million years. Evolutionary biologist Neil Shubin points this out in his book *Your Inner Fish,* where contingency is the major theme of vertebrate evolution. At the same time, we cannot oversimplify. It would be nice if we could oversimplify the structure of our brain; then we could explain it as a layering process, which neuroscientist Paul MacLean attempted with his triune brain hypothesis in the 1960s. By this hypothesis our inner brain is the product of early evolution of the vertebrates. This so-called lizard brain would control our basic functions such as detecting cold or heat, monitoring hormonal levels in our bodies, and monitoring blood glucose levels. This unemotional part of the brain would be our autopilot for fight-or-flight responses to environmental challenges.

MacLean called the next part of our brain the so-called mammalian brain, encompassing the limbic system. This is a complex system with many moving parts: the hypothalamus, the hippocampus, and the amygdala (to name a few regions), all surrounded by the cingulate gyrus. This part of the brain can also be divided into two subparts called the paleolimbic region and the neolimbic cortex. The paleolimbic is responsible for mediating social behaviors and hierarchy—somewhat primitive behaviors when it comes to mammals, essential for group survival and harmony. The neolimbic cortex is a bit more derived, because it is responsible for organizing experiences and memories. The final region in MacLean's layered way of thinking about the brain is the prefrontal cortex or the "human" part, where planning and thinking occur.

Is Your Brain "an Onion with a Tiny Reptile Inside"?

All of this is not quite as accurate as we would like. In fact, we have known for decades that it only comes close to explaining the full complexity of the human brain's structure. Why should we discard the triune brain? If we take it literally (and in science we have to take things literally when trying to explain things), then it does not accurately describe how the neural tissue of our brains evolved. According to psychologists Joseph Cesario, David Johnson, and Heather Eisthen, there are three major reasons to abandon the triune brain. First, it resurrects a discarded idea about nature called *scala naturae* (nature's ladder). This idea posited that organisms arise on an increasing complexity ladder, with lower species changing into species on a rung of the ladder that is higher. It all culminates with humans on the top rung of the ladder and no rungs above. (Early Christian theologists implied that two other rungs exist above humans: one for angels and one for God.) *Scala naturae* has been repeatedly rejected by modern evolutionary biology. One living thing does not evolve or change into another living thing in nature. Instead, ancestors exist that have common characteristics of their descendants, which natural selection can use to mold the descendants. Common ancestors allow for different lineages to evolve similar structures. For instance, cephalopods have rather complex brains, with complex behaviors that go with them. But their brains are not the same as ours. As with trichomes, something that is successful can evolve independently many times and in many different ways. *Scala naturae* and the triune brain imply only one outcome: something more complex or more perfect.

The second reason Cesario and colleagues cite for the faultiness of the triune hypothesis is that it assumes that complexity of the brain is additive: nature adds a little prefrontal cortex here, and the organism gets a little smarter. It adds a lot there, and the organism gets really smart. But we have known for a long time that larger neural real estate does not necessarily mean more complex structures or behaviors. It is how the organism uses added or expanded neural tissue to cope with the outside world that makes it more complex behaviorally.

The final argument from Cesario and colleagues is that the three or four layers of the triune brain were there at the very beginning. There is no reason to think they were added by layering. A reptile's brain is not simply made of a single layer of cerebellum and brain stem tissue that controls primitive functions. Nor is a primitive mammal brain simply a limbic system layered over the tissue of the brain stem and cerebellum. The early mammalian brain also had the other layers of the fully developed brain. And humans do not simply layer a prefrontal cortex over the limbic system. The difference is that more of the existing neural real estate is dedicated to thinking and decision making in the human brain than in a reptile brain. So the number of layers does not dictate complexity (behavioral or structural), but rather how those already existing layers have been altered over evolutionary time and how they connect to each other. The title of Cesario and colleagues' paper—"Your Brain Is Not an Onion with a Tiny Reptile Inside"—speaks volumes as to the inadequacy of the triune brain hypothesis.

Many psychology and biology textbooks, however, still use MacLean's triune brain to explain brain structure, even though it is not fully accurate. It is a handy heuristic for getting one's toes wet when studying the human brain. If you understand the concept of the triune brain, you have a good starting point for grasping the overall complexity of our nervous system, because the tenets of the triune brain can all be tested—and though they fail the tests, they suggest which questions to ask or which hypotheses to test further. The bottom line is that as organisms evolved, their brains changed by altering the size and function of specific regions. Humans ended up with a rather complex organ, but other organisms evolved equally complex brains.

This story of the triune brain has introduced some of the major segments that factor in how marijuana impacts our brains. Let us consider how marijuana and THC in particular impact neural processing. There are a huge number of cells in the human brain—an estimated 86 to 100 billion—and most of these need to communicate with each other. Not all brain cells communicate with each other the same way. What's most relevant to cannabis is how nerve cells communicate by combining electrical and chemical signals, which can move from cell to cell in the body and the brain. This complex system of "wiring" is facilitated by connections called synapses, and the typical neuron can have hundreds of thousands of connections from itself or other neurons through these synapses. There are also other kinds of cells in the brain, spinal cord, and all of the rest of the peripheral nervous system. These cells do not create action potentials and are called glial cells. Their main function is to support and protect the neural cells that generate electrical impulses and act as housekeepers for certain neurotransmitters. Because the chemicals in cannabis work mostly in the synapses, we need to delve into these intricate signaling switchboards.

Synapses

The development of any animal starts from a single fertilized egg cell and results in a fully mature organism. Although there are only about a dozen major kinds of cells in a mature human body (for example: stem, muscle, nerve, sex, and skin) there are over 200 specific kinds of cells into which these major cell types can differentiate. Contrast this with about ten mature cell types in sponges, or about six in placozoa, the pancake organism described in chapter 3. Plants have only three major kinds of cells, but these can differentiate into a large number of mature cell kinds (such as trichomes).

Nerve cells in bilateral animals like us humans are quite complex, varying in function, shape, and size. There are three major kinds of functional nerve cells— sensory, motor, and interneuronic. Figure 10.1 shows a highly schematized rendering of these cell types in a neural reflex system. Sensory nerve cells receive some sort of external stimulation, which is passed on to other cells in the system along axons, synapses, and cell bodies. Interneurons pass the external information along the nervous system and to the brain again through axons, cell bodies, and synapses. The motor cell enervates a response such as a muscle twitch through the cell bodies, axons, and synapses, leading to the region of the initial stimulus.

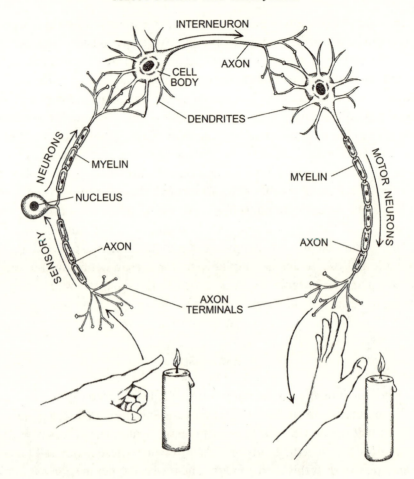

Figure 10.1. Three major kinds of nerve cells at the functional level—sensory, inter-
neuron, and motor. How these cells communicate with each other is also pictured.

As for cell shape, an introduction to nerve cell structure is warranted. Some
neural cells are thought to be quite long. The motor neurons that enervate the
lower part of the human body can be up to four feet long. These cells run from
the end of the spinal column at about the hip to the tips of the toes. These cells
(longest in the human body) carry information to and from the distal parts of
the body to the brain. The basic structure of a nerve cell consists of the cell body,
the axons, and the dendrites. The cell body is where the cell nucleus is and where
DNA coding for gene products resides. Axons are long protrusions of the neuron

cell, along which the electrochemical signals are carried. Figure 10.1 shows the axons of sensory and motor cells coated with myelin, which serves as "insulation" for this part of the cell and provides for the smooth transmission of electrochemical signals.

There are three major ways that these cells can connect with one other. One way is from the tips of the axon (called terminals) to dendrites from other cells. This is called an axodendritic connection. This dendrite—axon terminal link is not truly a connection, but rather a small open space (a synapse) that uses chemical exchanges to implement electrochemical signals called action potentials. The second kind of connection, called axosomatic, links the axon terminal with the cell (soma) body itself. The final connection, called axoaxonic, describes axon–axon exchanges.

There are two sides to every synapse—a pre-side and a post-side. On the presynaptic side is a cell where a signal exists that needs to be transmitted to the postsynaptic side. Any single neural cell can be both presynaptic and postsynaptic at the same time, which indicates that a lot of information can be trafficked through any single nerve cell and in different directions. As noted previously, nerve cells can communicate via chemicals and electrical impulses. Most nerve cells in animals work by transmitting chemicals, so I will focus on this kind of transmission. The chemicals secreted across a synapse, called neurotransmitters, carry messages the brain ultimately interprets. The specific region of the brain where the chemical secretions occur will partly determine the neural message and response. And depending on the chemistry of the molecules being transmitted in the synapse, the message can be excitatory or inhibitory, meaning that the synapse can speed up or slow down the transmission of information. Understanding the major kinds of chemicals that are important in animal nervous systems will help us understand marijuana's impact on the human nervous system. I list some here (table 10.1 and fig. 10.2), but we can eventually narrow these all down to a few chemicals that are relevant to cannabis activity in the brain.

Neurotransmission is best viewed as a series of steps, even though it is much more complex than a simple recipe. The steps include (1) synthesis of the neurotransmitter receptors or obtention by your cells through ingestion; (2) release from the presynaptic neuron to the opening between the pre- and postsynaptic cells; (3) activation of neurotransmitter receptors on the postsynaptic cell surface; and (4) termination—the signal must terminate so that the synapse can reset itself for more information. All along these steps, the message is transmitted through what is called "action potential," a kind of electrical impulse.

Table 10.1. SOME COMMON ENDOGENOUS NEUROTRANSMITTERS

Neurotransmitter	Function	Inhibitory/excitatory
Acetylcholine	Learning	Excitatory
Endorphins	Euphoria	Excitatory
Serotonin (5-HT)	Moods	Inhibitory
Gamma-aminobutyric acid (GABA)	Calming	Inhibitory
Glutamate	Memory	Excitatory
Catecholamines	Stress	Excitatory
Adrenaline (epinephrine)	Fight or flight	Excitatory
Dopamine	Pleasure	Both
Cannabinoids	Anti-inflammation, pleasure	Both
Anandamide	Pleasure	Both

STRUCTURES OF NEUROTRANSMITTERS

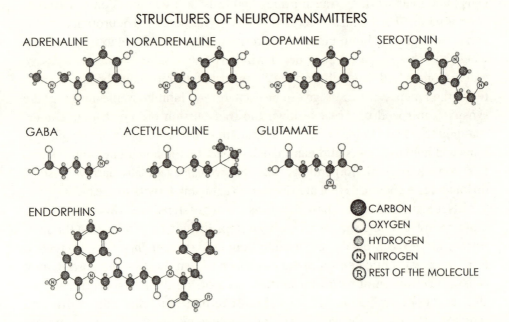

Figure 10.2. Structures of neurotransmitters.

Release

There is a lot of electrochemical activity in your cells that is responsible for neurotransmission. Many small molecules (including ionic forms of calcium and potassium) course around neural cells, floating in and between the nerve cells. The ionic forms of these atoms attempt to complete themselves by binding to some other atoms, so that they can easily move about the cell and the synapse. These ionic forms are charged (either positively or negatively), and certain movements of these ions carry electrical charges—the action potential mentioned previously. The ions cannot simply cross cell membranes; they need to be invited into the cell or escorted out via proteins that make what are called gated ion channels. These small gates or pores can open and close, depending on the ionic milieu in the synaptic cleft or on whether or not they have been activated by a protein inside the cell. Basically, neural activity is the result of all the involved cells' attempts to regulate and keep track of the ionic charges moving inside and out of synapses. If there is too much of one ion in a synaptic cleft, a gate will open to allow the ions to escape into the postsynaptic cell, which changes the charge in the postsynaptic cell. The change in the postsynaptic cell then travels through the nerve cell to the next synapse, or ends and triggers a response in the brain.

Where do these small molecules originate? Some are synthesized from the genes of your genome, and others find their way into your nervous system after being ingested in some way. Once ingested or synthesized, the small compounds get into your bloodstream and move to the brain, where they are packaged into large blob-like bodies called vesicles and transported to the synapse down the axon of the nerve cell. When the vesicles get to the synapse, they are then ready to act in the synaptic clefts. When the blobs reach the presynaptic cell surface at the synapse, they are ready to spew out their contents into the synaptic cleft—but first they need to be signaled to release their load. This signaling is accomplished by the action potential moving down the dendrite of the cell toward the synapse.

The action potential destabilizes the cell membrane, which responds by opening the gated channels, allowing for the influx of other ions into the cell. This complex release involves electrochemical changes in the membrane of the postsynaptic cell. The mechanism also involves proteins made by your cells called soluble N-ethylmaleimide-sensitive-factor attachment protein receptors. Neurobiologists have termed these modulatory proteins SNAREs. These proteins help the vesicle that was attached to the presynaptic cell membrane and eventually induce the release of the neurotransmitters into the synaptic cleft, where they interact with the postsynaptic cell membrane. And so on, until the stimulation is over and the nerve cells need to reset themselves. But how do the small neurotransmitters get into the cells of the synapse and signal those cells to carry the signal onward?

Signaling

After the neurotransmitters are released into the synaptic cleft, one of two types of receptor proteins on the cell surface will bind to the small neurotransmitter molecule. These two types of receptors are the gated ion channels mentioned above, and a category of important signaling proteins called G-coupled protein receptors (GPCRs). Both kinds of proteins are made by your cells from genes encoding them in your genome. The gated channels work extremely fast (on the scale of milliseconds) and can either excite the cell to communicate or inhibit further communication. The GPCRs are a big family of proteins that are threaded through the membranes of cells with a receptor site on the outside and a response mechanism on the inside. They act relatively slowly (on the order of seconds or minutes), because the mechanism is more complex and involves more steps than simply swinging a gate. Once a GPCR binds to a neurotransmitter, it changes its shape both inside and outside the cell. This change on the inside triggers other compounds inside the cell (called secondary messengers) to bind to the receptor. The binding produces an interior signal that can be transmitted further or terminated. These extra steps in the secondary messenger system are called a cascade, and it is the cascade that slows down the response of the GPCR system in the nervous system.

Reset

With all of this swapping of ions and activation of proteins, there is a lot of clutter produced in the synapse that if not cleaned up will cause the nerve cells to be incapable of further functionality. Something needs to happen to reset the system for further transmission: a sort of housekeeper for the synapse is needed. There are three ways to inactivate the transmitted signal. The first and simplest is through diffusion of the ions and neurotransmitters out of the synaptic cleft and into the bloodstream, where they can be recycled or destroyed. As the molecules diffuse from the synaptic cleft, the concentration will lower and gradually tone down the postsynaptic response to the neurotransmitter. This effectively resets the synapse for further interaction with other neurotransmitters.

A second way the synapse cleans up is through enzymatic destruction. Enzymes such as monoamine oxidase (MAO) and catechol-O-methyltransferase (COMT) are produced by cells (from the genes for them encoded in your genome), seeping into the synaptic cleft and destroying any neurotransmitters present. These enzymes are specific for certain neurotransmitters. There are two MAOs (A and B). MAO-A breaks down serotonin, norepinephrine, and epinephrine. MAO-B and MAO-A both break down dopamine, and MAO-B attacks some specific small neurotransmitter molecules. COMT is responsible for breaking down catecholamines including dopamine, epinephrine, and norepinephrine. Endocannabinoids are degraded by two enzymes: monoacylglycerol lipase (MAGL) and fatty acid amide hydrolase (FAAH). Exogenous plant cannabinoids ingested by humans are degraded by an enzyme called cytochrome P450. The MAGL molecules will clear the endogenous cannabinoids your body makes, and the FAAH molecules will clear plant cannabinoids that have gotten to your brain.

The final way housekeeping is maintained is through a cellular process called reuptake. Nerve cells themselves can accomplish reuptake. Both pre- and postsynaptic nerve cells use a "blebbing" process called endocytosis to clean up the synapse. (Blebbing somewhat resembles pushing a finger into a balloon.) At the cellular level those indentations can then fill with molecules slated for reuptake, seal over the outside of the bleb, and deliver the molecules to the interior of the cell. This process efficiently delivers the neurotransmitters back to the pre- and postsynaptic cells that can reuse them. Reuptake thus makes it easier on the cell to provide important neurotransmitters over and over again without having to make more each time a neural response is needed.

Another way reuptake can be accomplished is through the action of glial cells. One important neurotransmitter that benefits from this kind of reuptake is glutamate. This small molecule is released into the synapse during some neural activities. Cells are susceptible to the toxicity of glutamate, so an enzyme converts the glutamate to a similar molecule called glutamine. Glutamine is neither toxic nor active as a neurotransmitter. As noted previously, glial cells, through the action of proteins called transporters, can collect glutamate from the synapse, convert it to glutamine, and transport it back to the neuron to be used again, creating an action potential in the postsynaptic cell.

There is a lot going on here. *Many* of these cell-to-cell connections can be made in your nervous system. Each nerve cell in your body can have thousands of dendrites stretching out of it and connecting to other nerve cells. There are over 100 trillion (that's a one with fourteen zeros after it) potential cell connections that can be made in your nervous system. All of them are chemically communicating with each other, and the sum of their communication is your behavior or your response to stimulation from the outer world. Molecular biologist Francis Crick, the codiscoverer of the structure of DNA, poetically characterized these processes: "A person's mental activities are entirely due to the behavior of nerve cells, glial cells, and the atoms, ions, and molecules that make them up and influence them." This is indeed an amazing interpretation, but one that is marvelous in its complexity. Endocannabinoids exist to regulate this astonishing system so that we respond to physiological stimulus in a way that is beneficial to us. Exogenous cannabinoids, such as those found in marijuana, can trick this complex, clumsy, beautiful system and alter the way our "nerve cells, glial cells, and the atoms, ions, and molecules that make them up" work—sometimes resulting in pain relief or a euphoric feeling.

11

Brain Science, Bliss, and the Endocannabinoid System

For centuries people who have studied the human body have recognized that we are complex organisms (more so than some other organisms and less so than others). As with many big ideas physical or philosophical, Aristotle comes into play when we consider the complexity of organisms. In *De Partibus Animalium* he offered a scheme for understanding the what, how, and why of the various parts of animals. He not only attempted to describe the anatomy and physiology of animals but also delved into whether the forms were designed by chance or by some divine intervention.

Describing the head and brain, he wrote, "The head exists mainly for the sake of the brain." So far, so good. Our heads and skulls are effective brain containers. But for Aristotle the brain was a radiator for the heart, where real sensing activity resided. He did recognize the head as the seat of some sensation, but overall he believed that "the brain, then, tempers the heat and seething of the heart." Whether or not he was wrong about the brain, though, Aristotle achieved something wonderful in *De Partibus Animalium.* He created what some scholars call "a comparative organography." He established that bodies were made of organs, which in general presided over specific physiological activities and functions.

As with most original thinking in Western humans, there was not much progress beyond Aristotle and his students. Except for some novel and important work in the Arab world and the Far East, a sort of conceptual dead zone occurred for almost two millennia until a great age of anatomical discovery and analysis started at the end of the 1700s and early 1800s, when organs and organ function were explored extensively. At the end of the eighteenth century, Félix Vicq d'Azyr, a French medical specialist, suggested there were nine general properties of life. These included digestion, circulation, breathing, and reproduction, which are now associated with our digestive system, our circulatory system, our respiratory system, and our reproductive system. Over the next few centuries, other organ systems were added to the list. Perhaps two of the most important to be added during this time were the nervous system and the immune system. Study of these two systems led to a deeper understanding of the molecular phenomena involved in body systems.

Counting organs, organ systems, and body systems might seem to be an esoteric task. But oddly, scientists have settled on ten or eleven major organ systems in the human body. These systems are skeletal, muscular, nervous, endocrine, cardiovascular, lymphatic, respiratory, digestive, urinary, and reproductive. This list ignores skin (the eleventh system in an eleven-system scheme). Given that our exterior integument is one of the largest organs in our bodies, this could be viewed as a major oversight, but it also points to the difficulty of recognizing the various parts of our bodies as integrated systems.

There is, however, one further system that has been a latecomer to the fray. The synthesis of THC in the 1960s led to an entirely new area of brain science in human biology. It was inevitable that discovery of this last system occurred so late. THC was the trigger for the discovery of this system, and research into an entire body system was stymied because of the illegal status of cannabis. The story of the discovery of this system actually starts with another category of controlled substances: the opioids. In 1973 researchers determined that the small opioid molecules such as heroin and morphine would bind to specific areas of the brain. Johns Hopkins University researchers discovered that there were receptors for these opioids in the brain. Making the leap from opioid receptors to receptors for the active ingredient in marijuana was simple. But it took fifteen years to document their existence conclusively, when researchers at St. Louis University School of Medicine showed that the brain had receptor molecules embedded in the membranes of nerve cells that recognize analogs of THC. Allyn Howlett and William Devane were the scientists who pinned this down, and they further showed that these receptors were much more ubiquitous in the brain than any of the other receptors previously examined, such as acetylcholine, endorphin, serotonin, glutamate, or gamma-aminobutyric acid (GABA) receptors.

The next question was where in the body these receptors resided. Researchers knew that the brain harbored many of them, but suspected the receptors for THC were also elsewhere in the body. In the 1980s the Pfizer pharmaceutical company was able to synthesize an efficient THC analog that could be labeled with radioactivity and injected into the body. By detecting where the radioactivity ended up, presumably bound to natural receptors, researchers could also determine where the receptors for this cannabinoid existed. To their surprise the receptor sites were distributed throughout the body, making this system more widespread than just the brain. In 1990 pharmacologist Lisa Matsuda and her colleagues announced the cloning and sequencing of the gene for cannabinoid receptor molecule from a rat brain. Thirty years after the chemical characterization of THC, researchers had pinpointed the receptor for this small cannabinoid molecule. The receptor molecule Matsuda and colleagues characterized was called the CB receptor, and its sequence was similar to a category of proteins being rapidly discovered at the time called G-coupled protein receptors (GCPRs). They are all recognizable because they reside in cell membranes and have seven highly characteristic helices that wind back and forth through the membrane, with one end of the protein extending into the cell and one end sticking outside. Three years later molecular biologist Sean Munro and colleagues cloned and characterized another molecule that served as a receptor found in the spleen. Since this was the second CB receptor found, the original CB receptor became CB-1 and the new one was named CB-2.

The story wasn't yet complete, however. Researchers had found the receptors that bound THC, an exogenous cannabinoid. Certainly if these receptors existed, then they had to have some interaction with an endogenous molecule or multiple molecules that were part of the body's normal functioning. In 1992 a group of pioneering researchers isolated and characterized this endogenous factor: an endocannabinoid that they named anandamide (AEA). Three years later they discovered and characterized a second endocannabinoid called 2-arachidonoylglycerol, or 2-AG. The differences between AEA and 2-AG are discussed in detail below. Because of its ubiquity in the human body and the importance of its function, this receptor—ligand system was named the "endocannabinoid system."

"I Have Never Used It"

How do we know so much about cannabinoids and the endocannabinoid system? Perhaps primarily because of one man: Raphael Mechoulam. During the writing of this book, this amazing scientist passed away at the age of ninety-two. He remained active as a scientist until his death on March 9, 2023. Some call him the father of cannabinoid/endocannabinoid research; he is best known for discovering THC in 1963 and the primary endocannabinoid of our brains—anandamide—in 1990. ("Discovering" means the first to isolate the compound and determine its structure and chemistry.) His early life included hiding from Bulgarian Nazis during World War II and immigration in 1949 to Israel, where he did outstanding chemistry research. When he was appointed to his first professorship in the early 1960s at the Hebrew University of Jerusalem, he scoured the literature for a study system that would offer long-term research opportunities. Being trained in the chemistry of natural compounds, he focused on plant systems where he might find a compound that could be used in chemistry or pharmaceutical development. His literature search unveiled a strange observation. Despite its important role in plant phytochemistry and in illicit drug trade, the active chemicals of cannabis had little to no research focused on them. Here was the intellectual gold mine he was seeking. But he quickly realized that the reason for the lack of research on cannabis was probably the illicit, illegal, and punishable aspect of possessing cannabis. Like any clever person, he found a work-around, as detailed in his 2023 memoir:

> How does one get cannabis—a strictly regulated illicit drug—in sufficient amounts to initiate research? In 1963! Again, I was lucky. The administrative head of my Institute knew a police officer, who was presumably the number two (or possibly the number three) in the Israeli Police hierarchy. He phoned and told him that a Dr. Raphael Mechoulam needed hashish for research and that he—meaning me—was completely reliable (though he barely knew me). I just went to Police headquarters, had a cup of coffee with the policeman in charge of the storage of illicit drugs, and got 5 kg of confiscated hashish, presumably smuggled from Lebanon.

Given that another famous scientist, neurologist Oliver Sacks, had a penchant for taking and experiencing his research drugs, I wondered whether Mechoulam

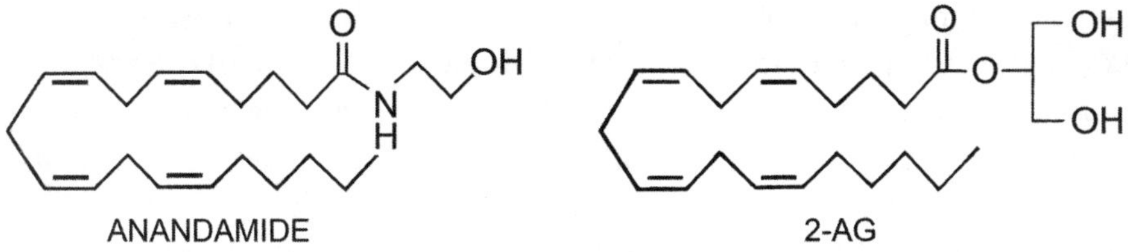

Figure 11.1. Structure of anandamide and 2-AG.

partook of the noble weed. In an interview for *Culture* magazine in 2017, he said, "I have never used it." This was after fifty-five years of research on the plant and its psychoactive effects. Mechoulam pointed out that any departure from being completely legal about using the hashish would have landed him in jail and, worse, would have destroyed his chances of continuing work on the amazing plant.

His pioneering work on the endogenous cannabinoid called anandamide (from the Sanskrit word *ananda,* meaning "bliss") led to the better understanding of how our bodies make large amounts of cannabinoids that move throughout our endocannabinoid system. Anandamide (AEA) exerts an overall modulatory effect on the brain reward circuitry and can bind to endocannabinoid receptors throughout the rest of the body to suppress pain and tumors. Another endogenous cannabinoid is 2-arachidonoylglycerol or 2-AG, which is found in higher concentrations in the endocannabinoid system than AEA. There are other endocannabinoids such as noladin ether, palmitoylethanolamide (PEA), virodhamine, and oleoylethanolamide (OEA) that our bodies make, but AEA and 2-AG are the major players in the endocannabinoid system. Figure 11.1 shows the chemical structure of these two major endocannabinoids.

The difference between anandamide and 2-AG are their "tails." To me these resemble the "centipede" triterpenes discussed in chapter 9. They are about the same length and have a similar look as the exogenous cannabinoids THC and CBD. Even a slight resemblance of an exogenous molecule to another active endogenous one will cause confusion in the function of the endogenous one.

Where do we get these important neurotransmitters? They are synthesized from genes in our genomes, and our bodies make a whole slew of enzymes that regulate the production, storage, and release of these chemicals as we respond to physiological challenges. The system of endocannabinoid synthesis and maintenance is quite complex, involving tens of different enzymes and proteins. It is in and of itself a mini-kluge, as these proteins work in concert in convoluted ways to regulate the cellular activity of endocannabinoids. A diagram of the endocannabinoid system in a cell might make sense to someone working on the system, but to the lay eye it would look much like the *Mouse Trap* game I played as a kid. Different parts of the cell have different biochemical contraptions that take care of a step or two of the inner workings of the endocannabinoid system. They are all connected and work together to regulate the endocannabinoid system.

So far, we have examined two of the three major parts of the endocannabinoid system (endocannabinoids and where they bind). The third part is perhaps the most interesting: the receptors that recognize the endocannabinoid molecules and transfer the recognition into information in nerve cells. There are two major receptors for endocannabinoids, conveniently named CB-1 and CB-2. These two receptors reside in the cell membrane of a cannabinoid-sensitive neuron. Since they are GPCRs, part of their structure sticks out of the cell and into the synapse (if one is present). On the inside end of these kinds of GPCRs is a structure that binds other proteins as described earlier. When a cannabinoid receptor site on the synapse side of the membrane encounters something it will bind to, it does so and produces a change in the overall structure of the GPCR, which then produces an interaction with intracellular protein complexes and induces signaling.

If you look closely at these two structures, you will see well-defined similarities. Each of the two proteins has seven helical stretches. The placement of the helical structures is in the same general positions in both proteins (fig. 11.2). There are some slight differences between the two proteins, though, which make them behave differently when confronted with endocannabinoids and exogenous cannabinoids. These structural differences make the specificity of binding to and the responses of CB-1 different from the responses of CB-2. And where these receptors are synthesized in our bodies makes all the difference as to how these two receptors behave. Figure 11.3 shows a diagram of the human body and where the two receptors predominate.

Although the position in the body of these two receptors overlaps considerably, there is a general trend in how they are dispersed. CB-1 appears to be very brain centralized in its distribution. It also appears to follow general tracts of the peripheral nervous system. CB-2 on the other hand is truly widespread in the body but does show centralization in peripheral organ systems primarily in locations where immune function is important. As the figure implies the two receptors have very different bodily domains of effect.

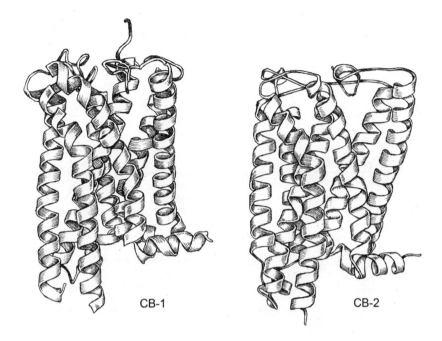

Figure 11.2. Ribbon models of CB-1 (left) and CB-2 (right). Note that there are seven helical parts of each protein (these will span the membrane of the nerve cell) and out-of-cell and inside-the-cell ends (at top and bottom of each figure). Adapted from Shao et al. (2016).

Evolutionary Shuffle

When did these receptors first arise, and how widely dispersed are they in the tree of life? Because the endocannabinoid system is so important in humans, knowing where and when it originated in animal evolution constitutes an essential cog in understanding their function. One might expect the origin of the endocannabinoid system in animals and the origin of cannabinoid synthesis in plants to be correlated. At least that would be a good hypothesis to test. A good way to perform this experiment is to identify the genes in the genomes of as many animals as possible for the cannabinoid receptors CB-1 and CB-2, and then to investigate how those changed with time. This kind of approach can also determine the animal group (or common ancestor) from which certain genes were gained.

But this approach also begs the question of how a new gene is acquired.

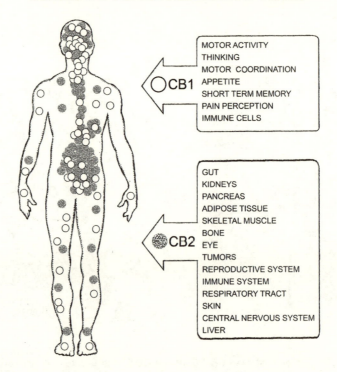

Figure 11.3. Locations of CB receptors and their various functions. Adapted from Muralidhar, Maurya, and Velmurugan (2019).

Some new genes are obtained by organisms through horizontal transfer or "jumping" from one species to another. This mode is probably not how most animals and plants make new genes, but it does predominate in bacteria and archaea. Another mode is to simply duplicate an existing gene and let natural selection mold the duplicated gene into something quite different. This mode is a tenable one for the CB genes. Yet another mode is more drastic, but it has been shown pivotal in animal genome evolution. This mode involves duplicating the entire genome of an organism during reproduction, so that instead of having four chromosomes for instance in the diploid stage, the organism now has eight. The genes on the duplicated four chromosomes are then "free" to diverge and attain novel functions. This mode is particularly tenable for accruing new genes for animals and plants; researchers have determined that full-scale genome duplications have occurred in the evolution of several eukaryotic lineages. Multiple genome duplications have occurred in plants and fish. One specific entire genome duplication that has been identified in the common ancestor of vertebrates has been touted as an important source for novel genes that led to the diversification of vertebrates.

John McPartland, along with his colleagues Isabel Matias, Vincenzo Di Marzo, and Michelle Glass, tested this hypothesis one step better in 2006. They examined not only the distribution of CB-1 and CB-2 in the animal tree of life, but also an array of other gene products involved in the synthesis of AEA and 2AG endocannabinoids. These are the enzymes that make up the endocannabinoid system mini-kluge described earlier. McPartland and colleagues' studies were undertaken when there were only about a dozen full genomes of organisms that had been sequenced. McPartland and his colleagues searched the sequence database of each of these twelve genomes (a human, a mouse, a tunicate, an apicomplexan, a ciliate, a fish, an insect, a nematode, a fungus, a plant, an archaean, and a bacterium) for the genes that code for ten of the proteins essential to the endocannabinoid system. Once these genes were found, a gene family for each one could be estimated using phylogenetic methods. By examining the family trees, one could then determine which groups have which genes.

The collection of twelve species in McPartland's analyses is not a perfect representation for a complete tree of life (many major groups are missing), but it will suffice to trace the working parts of the mini-kluge that the endocannabinoid system eventually became. Because this kind of research is what I do for a living, I have spent some time repeating this experiment with the thousands of organismal genomes that are now available; McPartland and his colleagues got it largely right working with just these twelve species. Figure 11.4 shows a tree of life and the origin of each of the ten genes used by McPartland and colleagues.

It should be obvious from figure 11.4 that the endocannabinoid system did not just pop out of thin air, fully functional. If it had, then all the alphabet soup names of enzymes would appear on only one branch of the diagram. Instead the endocannabinoid system was molded in steps over millions if not billions of years of the evolutionary process. For instance, the fatty acid amide hydrolase (FAAH) gene that catabolizes AEA arose in the earliest of eukaryotes. But wait: there probably was no AEA being produced in these early eukaryotes that lived about 1 to 2 billion years ago, because the enzyme responsible for synthesizing AEA did not arise until the ancestor of fungi and animals. (Yes, fungi and animals have a

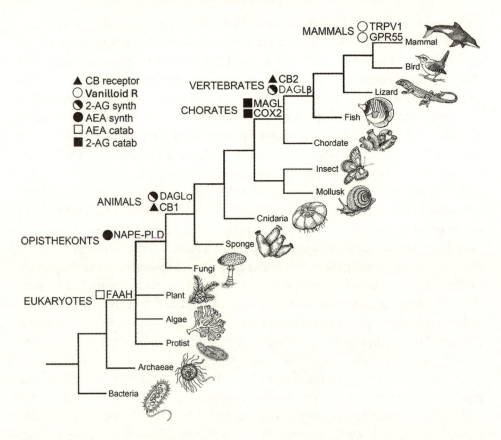

Figure 11.4. Tree of life showing where the ten proteins arose in the endocannabinoid system.

more recent common ancestor than plants and fungi; in other words, a mushroom is more closely related to us than it is to a sunflower.) Almost assuredly, fatty acid amide hydrolase arose as part of a completely unrelated processing pathway and was "co-opted" by the AEA pathway. Another example of the stepwise construction of the AEA pathway is the production of 2-AG. This part of the pathway probably arose as a two-step process with DAGLα arising in the common ancestor of all animals, and DAGLβ arising in the common ancestor of all vertebrates. A final example concerns the finishing touches in the endocannabinoid system. Remember that AEA is a neurotransmitter that bonds to other receptors. So what it interacts with becomes an important part of the endocannabinoid system. One of these receptor molecules is called the vanilloid receptor (TRPV1), and the second is a receptor called GPR55. The genes responsible for synthesis of these receptors arose in the common ancestor of mammals, indicating that once the full endocannabinoid system came into existence in the common ancestor of vertebrates, the story of neurotransmission wasn't over. The various steps to the endocannabinoid system happened at radically different times, with somewhat large gaps of time between the steps.

A Delightful Trip

Let's go on a "delightful trip" (Mechoulam used this phrase to describe his career in endocannabinoid system study) and follow one endogenous cannabinoid molecule and one exogenous cannabinoid molecule through a neural synapse. Let's start with the endogenous cannabinoid AEA, which first needs to be synthesized by our cells. A clumsily named molecule NAPE-PLD (N-acyl phosphatidylethanolamine-specific phospholipase D) synthesizes AEA from starting material called arachidonic acid. We mostly obtain this precursor molecule from foods such as meat and eggs, and our cells sequester it for use in synthesizing molecules that are important to the endocannabinoid system like anandamide (AEA). The chemistry is rather complex, as our bodies can also synthesize arachidonic acid, but the most reliable source of this molecule is through our diet. Once the AEA is synthesized, it remains in the cell awaiting its use in the endocannabinoid system. As we explore the effects of AEA on our nervous system, we need to keep in mind that it is very fragile compared with THC and CBD. It breaks down easily into arachidonic acid and other components, which is the job of the fatty acid amide hydrolase enzyme shown in figure 11.4. THC and CBD are tougher molecules and like to linger for longer periods, especially in fatty tissues.

The synthesized AEA waits in the cells for some physical or physiological stimulus to go to work. As with the endocannabinoid system, AEA also has many neurotransmitter functions, including regulation of sleep, memory, appetite, pain regulation, and hormonal regulation of ovulation in females. The other major endocannabinoid, 2-AG, is also synthesized from arachidonic acid using different enzymes than does AEA. It is more abundant in our cells than AEA and is the major neurotransmitter that interacts with CB-2. Both AEA and 2-AG will interact with CB-1.

Most transmission of information via our synapses is routed from the presynaptic cells to the postsynaptic cells. But not AEA and 2-AG in the endocannabinoid system: their movement is reversed. AEA and 2-AG are synthesized primarily in the postsynaptic cell, and when induced these neurotransmitters will travel backward across the synaptic cleft and bind to CB-1 and CB-2 receptor proteins in the membrane of the presynaptic cell. By doing this they can regulate the strength of the synaptic connection (i.e., how efficiently it transmits information) and hence the long-term utility and function of the synapse. This process, called retrograde signaling, affords an impacted synapse a lot of versatility and enables the AEA and 2-AG neurotransmitters to be both excitatory and inhibitory. Since the AEA impacting part of the endocannabinoid system is distributed in different regions of the brain, AEA will have effects on the many different kinds of neural regulation systems such as the dopamine system or the serotonin system. In other words, it will regulate other neurotransmitters and either tone down or ramp up their activity.

Let's return to our AEA molecule, which has been synthesized in the postsynaptic cell and is ready to be released into the cleft of the synapse. When you are hungry, many different neurotransmitters are synthesized as a result. But let's also posit that as a result of smelling something really tasty, a signal from your olfactory system triggers the release of the newly synthesized AEA into the synaptic cleft in a region of your brain where there are dopamine-regulated synapses. Actually, this will be happening in many other brain cells as a result of the olfactory stimulus. The AEA travels backward across the synapse and finds a CB-1 or CB-2 receptor on the presynaptic cell which it binds to. Such binding triggers a regulatory response in the presynaptic cell that regulates GABA, which is the regulatory neurotransmitter that stimulates the release of dopamine. Dopamine is the pleasure molecule that drives many behaviors. When you are full, the synthesis of AEA in this part of the brain slows down, as does your appetite, because the CB-1 and CB-2 receptors are not triggered and the response is to stop downregulating GABA, which in turn downregulates dopamine. The rest of stopping eating is a balance between the reward system and other brain functions. The reward system is an important element in understanding THC use and abuse.

I have oversimplified a lot here. But the primary messages are that the endocannabinoid system evolved to regulate some of our behaviors, and that a lynchpin of this regulatory system is the synthesis and binding of our internal endocannabinoids (such as AEA and 2-AG) to cannabinoid receptors on many of our neurons (primarily CB-1 and CB-2). The simple ingestion of THC or CBD (or other natural and synthetic cannabinoids) can throw this regulatory system off in many ways, which is the reason why cannabis consumption can influence a

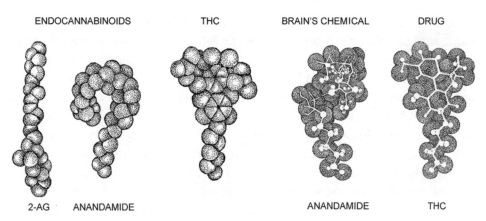

ENDOCANNABINOIDS   THC   BRAIN'S CHEMICAL   DRUG

2-AG   ANANDAMIDE   ANANDAMIDE   THC

Figure 11.5. Space-filling diagrams depicting the similarity of the shape and size of the endocanna-binoids and THC. The diagrams show carbons, hydrogens, and oxygens. Note that THC has a shape quite similar to anandamide.

variety of behaviors and physiological factors. This brings us back to following our THC molecule through the endocannabinoid system.

As explained in chapter 8, when you take a toke on a marijuana cigarette, much of the THCA in the cannabis is decarboxylated and rendered active by the heat of the burning cigarette. In the inhalant are billions of molecules of THC that enter your lungs. The majority of these molecules enter your bloodstream. Our THC molecule is one of these. It moves through the bloodstream to the brain, where it passes into the brain's neural tissue. The THC molecule makes it to a cluster of neural cell synapses that are part of the dopamine system. Due to the similarity of the endocannabinoids to THC (fig. 11.5), competition occurs for binding to the CB-1 and CB-2 receptors. Because there is so much THC that has been transferred to the brain via the toke, THC binds to many of the CB-1 receptors, clogging them up and stopping them from regulating GABA. This dysregulation enables a lot of GABA to be delivered to the synapse, and unusual loads of dopamine are pumped into the synapse. Hence the euphoric, pleasurable feeling one gets from smoking marijuana. The increased amount of dopamine also produces the effect of "wanting more" in your brain.

Elsewhere in your body after the toke, your bloodstream moves THC and CBD to all points in the endocannabinoid system, where CB-1 and CB-2 receptors reside. Both THC and CBD disrupt the proper functioning of the endocannabinoid system by binding to and clogging up the CB-1 and CB-2 receptors that are dispersed throughout your body. If some of the CBD molecules get into a tissue with cells that regulate pain, the many cells with receptors in this tissue can also respond to mechanical, thermal, chemical, or a combination of the three kinds of insult to the tissue. When a pain receptor cell is stimulated or irritated, it will communicate with nerve cells, which then send a message to the brain that some insult has occurred. Different pain receptors will elicit different responses that the brain interprets as heat, cold, or pressure. The alleviation of pain that THC and especially CBD deliver works in several ways regarding pain. One way is to simply block the nerve cells that transmit the insult from the pain receptor cell to the brain. A second way is to target receptors in the immune response and in turn reduce inflammation. Yet another way is to stimulate the production of adenosine in the brain, which has been shown to be an analgesic. Chapter 14 will delve into the many ways that CBD and other cannabinoid derivatives and synthetics can impact pain treatment.

The extent of the effects of THC and CBD relative to one another relies on the concentrations of both exogenous cannabinoids in the original intake toke. Consequently, adjusting the ratio of CBD to THC in many marijuana products has been an important but sometimes bewildering task. With the superhigh THC concentrated strains now available, the effects are much more powerful because they deliver more of the "clogging" agent to your neurons, and more of the neural effects that clogging produces. Like anything with cannabis, balance of effect is of the utmost importance.

12

Genes, Genomes, and Cannabis

Until about ten years ago, it was taboo to work with cannabis in the United States, even if you were an academic researcher you needed some kind of dispensation. And some major restrictions on the movement or transport of the plant hindered much of the research that could be done. If you were a researcher working on cannabis, you ran the risk that a representative of a legal or regulatory agency who had read your papers would come knocking on your door. That would put a damper on doing research on banned plants, much in the same way that Michael Pollan, author of *This Is Your Mind on Plants,* avoided being too specific about his experiences with the opium plants he grew in his backyard. But as research biologists Conor Jenkins and Ben Orsburn point out in their paper on generating the first cannabis genome, "The passage of the Farm Act of 2019 in the United States has allowed nation-wide access to *Cannabis* plants that possess a total concentration of THC of less than 0.3 percent weight. Coupled with de-criminalization and legalization of all *Cannabis* plants in Canada and an increasing number of states, the interest in *Cannabis* spans multiple areas of medical and industrial science."

In some ways it might have been for the best, although that research was held back at least for genomics, as it was only about ten years ago that DNA sequencing techniques moved into what was called the "next generation." The human genome and several other model system genomes were sequenced in the early 2000s. At that time, sequencing a genome the size of the human genome or the cannabis genome was a herculean task involving an estimated $3 billion and employing armies of technicians. The shift to next generation sequencing (NGS) accelerated the sequencing process by several orders of magnitude, and it also became several orders of magnitude cheaper to do.

Sequencing of the cannabis genome has resulted in unprecedented progress in the understanding of the plant's biology. Nearly all the genes involved in THCA and CBDA synthesis have been identified at the DNA sequence level, and the primary information in the genes in these pathways has been deduced. Other aspects of cannabis not related to THCA and CBDA such as growth time, flowering time, height, flower weight per total, and other phenotypes are also being studied using genome wide association studies (GWAS). The overall potential for understanding cannabis at the genetic and genomic level is poised to revolutionize cannabis cultivation, so it is important to understand the shift in resolution that we now have as a result of the next generation sequencing approaches. After pondering the impact of genomics on cannabis cultivation, Jeremy Plumb, who works at a commercial cannabis operation, recently predicted that "within three years none of the plants that we are growing currently will continue to be produced."

Some Genomics Fundamentals

The first thing to realize about cannabis genomics is that it is focused on the DNA sequence of the plant. At its most simple, the cannabis genome (like every genome of every organism on this planet) is a string of microscopic building blocks called nucleotides. These nucleotides come in four varieties in the genome of most organisms: guanine (G), adenine (A), cytosine (C), and thymine (T). That's all there is at the most basic level in the cells of living things (including viruses, if you count them as living). But there are millions of these nucleotides in linear arrangements, and it is the order of these nucleotides that imparts information to the cannabis cell.

How is the DNA packaged in the cells of the cannabis plant, and what does the packaging mean for its life cycle? There are two major ways cells are organized in organisms. The first way is simple: the DNA lives inside the cell membranes. The second involves packaging the DNA away in another membrane-bound structure called the nucleus. Microbes such as bacteria are examples of the former, and humans and cannabis are examples of the latter. The organisms with a nucleus surrounding their DNA are known as eukaryotes.

A sexually reproducing eukaryotic organism (like us and cannabis) has millions of cells that are basically of two types. The cells that make up most of a cannabis plant are diploid, meaning they have two copies of each gene in the genome. (These diploid cells are also known as somatic cells.) The other kinds of cells in a diploid organism are the gametes, which are haploid, with a single copy of the genome tucked away inside the cell's nucleus. When a pollen cell fertilizes an ovum, the amount of DNA returns to the diploid state and the nascent plant can start to develop. Diploid genomes such as that of cannabis hold tons of information in the guise of these Gs, As, Ts, and Cs, and a remarkably efficient mechanism has evolved to ensure that both parents contribute to the next generation.

The size of any genome gives us an upper limit of the potential amount of information contained in an organism's genetic makeup. The genome size can be determined using cytometric techniques (techniques that measure the components of cells), and it is fairly clear that diploid and haploid male cannabis cells have more DNA in them than in the corresponding cell types in females. It is a slight disparity, but nonetheless enough to recognize with available techniques. The difference arises from the way sex is determined in cannabis, which involves about 47 million nucleotides in number. This may sound like a lot, but it is only a 2 percent difference, as the male genome is 1 billion, 683 million nucleotides long, and the female genome is 1 billion, 636 million nucleotides long. The overall amount of nucleotides in the cannabis genome is average for a plant (fig. 12.1).

The DNA in the nuclei of eukaryotes is packaged further into smaller discrete bundles called chromosomes. Cannabis has ten of these bundles in each haploid cell, and ten pairs to make twenty total in each somatic cell. The cannabis genome is really a collection of linear stretches of nucleotides, as chromosomes are also linear strings of nucleotides. Chromosomes vary in size from organism to organism. For humans the twenty-three chromosomes of our haploid genome range in size from about 50 million nucleotides to 250 million nucleotides. The autosomal chromosomes in the human genome were named based on their size, as judged from microscope pictures (called karyotypes) of the genome showing the actual chromosomes in the cell. Chromosome 1 was judged to be the largest, and chromosome 21 the smallest. Sometimes a chromosome got its name based on this size criterion but later, when its size was determined more precisely with DNA sequencing, the convention was violated. This means that some human chromosomes do not fall in line with the largest-to-smallest naming convention.

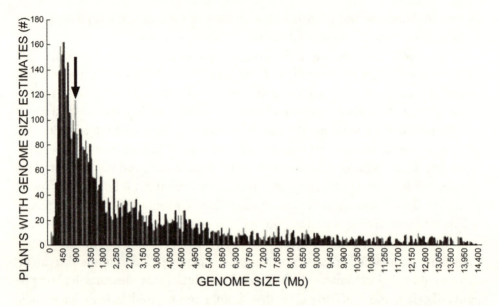

Figure 12.1. Distribution of the sizes of different plant genomes. Plant genomes range in size from about 63 million bases for *Genlisea margaretae,* a eudicot in the higher category of what are called asterids, or plants such as common daisies, to about 26 billion bases for *Sequoia sempervirens,* the sequoia tree. The arrow points to where the *Cannabis* genome lies in this distribution; it is neither very lean nor very obese with respect to genome size.

The same naming convention is conceptually true for cannabis (the largest chromosome is named chromosome 1 and so on to chromosome 10). As with human chromosomes, the size naming convention has been violated in some instances.

In humans, female cells have more DNA because of the sex chromosomes: namely, the X and Y chromosomes. Human sex is determined through an X–Y system, which is a parsimonious system that many organisms have adopted to ensure that both males and females are produced in each generation of the life cycle. Females end up with two X's (they are XX), and males are XY. The human X chromosomes carry over five times the number of nucleotides as the Y chromosome. Cannabis has a similar sex determination system called an X-to-autosome balance system, where sex chromosome makeup is XY for males and XX for females. Unlike humans, where the X chromosome is larger than the Y, the cannabis Y chromosome is bigger than the X, and thus XY individuals (cannabis males) have more DNA than XX individuals (cannabis females).

Whereas all the other chromosomes in the genomes of most eukaryotes come into contact with each other and exchange parts of their chromosomes with each other (a process called recombination), the Y chromosome and the X chromosome in an XY system recombine rarely. In humans, the Y chromosome is physically quite different from the X; it is smaller and more condensed. In cannabis, parts of the two sex chromosomes do recombine, but there is a long block on each of the two chromosomes that does not, and this is where the sex determining factors for cannabis lie. The rest of the chromosome is called "X-specific." This is a strange arrangement, and one that is very old in evolutionary terms.

It is strange because most plants are dioecious (hermaphroditic), with no need to have sex chromosomes. They only need the genetic instructions to make male and female genitalia. It appears that the XY sex determining system has arisen several times in the evolution of plants, and we can compare the sex determining systems of other plants to that in cannabis. When the XY system of cannabis and its relatives arose can then be estimated. It appears that the cannabis sex determination system arose at the very latest in the common ancestor of hops and marijuana over 20 million years ago, making it perhaps one of the oldest in plants.

If other members of the Cannabaceae (such as *Trema* and *Celtis*) have XY systems, then this would make the origin of the XY system in cannabis even older. Evolutionary biologist Djivan Prentout and colleagues found large amounts of divergence between the genes that are on both the cannabis X and Y chromosomes. This told them that the sequences of these genes are degenerating and will perhaps eventually be shed by the cannabis genome. This is a situation that is similar to many mammalian sex chromosomes, where it has been observed that the Y chromosome is degenerating and may someday be lost by mammalian genomes like ours. Apparently, cannabis is going through similar genomic growing pains as some animal systems.

Chromosomes are further subdivided into regions that code for gene products (usually proteins) and regions that do not (called intervening sequences). While these intervening sequences were at first thought to be extraneous and called "junk DNA," researchers are discovering more functional aspects of these regions that lie between the coding regions. There are two kinds of noncoding regions that come into contact with coding regions. The first type are the intervening regions between the coding region units. The second kind of noncoding regions are embedded within the coding regions, and these are called introns. The coding regions in these genes that are interrupted by introns are called exons. We humans have a little over 30,000 discrete genes in our genome, and cannabis has a little over 25,000 genes that code for gene products. These estimates are much lower than previously thought, as the human genome was initially thought to code for over 100,000 genes before the entire genome was sequenced. The reason for the discrepancy is that before the genome was sequenced, researchers were trying to estimate the number of different functions of genes that they thought would line up well with the overall number of genes. But it turned out that many genes can code for different functions, and the number of functions had been overestimated by about a factor of three to five.

Other information that a genome sequence can tell us includes how close in the genome certain genes are to each other (linkage); whether a gene has many copies of itself scattered throughout the genome (copy number); whether a gene has mutations relative to a reference genome; whether genomes have certain kinds of genes (like resistance genes or sex determining genes); and how variable a gene might be in a natural or agricultural population. All these factors are significant for the breeding and health of cannabis.

The way genomes are now sequenced relies on brute strength. That is, much more sequence is generated than what is contained in a typical genome. Let's posit that a genome has 1 billion nucleotides in it. One might need to obtain 100 times that amount to reach a full understanding of this genome. This amount of information would be called 100x coverage of the genome.

Sequences are obtained by fracturing the long genome strands into small fragments (usually a tiny fraction of the overall genome), and then these fragments are treated with biochemicals to produce the string of nucleotides on the fragments. When a genome is sequenced, it resembles a bunch of small pieces of DNA with strings of Gs, As, Ts, and Cs. To obtain the full genome from these short sequences, two processes need to be accomplished: assembly and annotation. Assembly involves taking the millions or billions of pieces of the initially fragmented genome of sequenced DNA and putting them back together. This is possible thanks to brute computational force. If the genome was initially fragmented using a random process, then every fragment will have some overlap with some other fragments. These overlaps can be used to lace the millions of small fragments together into what are called contigs (contiguous sequence fragments). It's like putting a linear puzzle together, and when you can't add any more to one contiguous reconstructed section, you start on a new one. Most of the time the breaks in contiguous reconstructions mean that the sequences come from different chromosomes (which is how the chromosomal structure of genomes can be inferred).

After assembly what you have are large strings of Gs, As, Ts, and Cs, which don't tell you much except maybe the number of nucleotides in the genome. This is where the second approach—annotation—comes in. Annotation involves the systematic translation of the DNA-based Gs, As, Ts, and Cs into the amino acid strings of proteins the DNA codes for. This is achieved by using a set of rules (the genetic code) for predicting what amino acid sequence a given DNA might produce. This is another computationally intense step, as every possible way of decoding the DNA sequence is accomplished and placed in a database. Then this temporary data base is searched using algorithms that determine whether the predicted amino acid sequences of the newly sequenced genome match with proteins known to exist in other organisms.

Amino acid sequences can fold into three-dimensional structures, which will render the protein some of its function. Thus another approach is to determine whether the predicted amino acid sequence makes a protein that folds into a shape that has a function. Both these approaches are used to annotate the DNA sequences so that we can ascertain which DNA stretches are potential genes and what the function of those genes might be.

Ten Chromosomes, 25,000 (or So) Genes, and 1,001 (or More) Secondary Compounds

Most of the genes in an organism's genome contain the basic instructions for building and maintaining the organism, and in this sense are not terribly exciting aspects of cannabis. These genes are shared across related organisms and are responsible for what make a eukaryote a eukaryote or a plant a plant, but they have little to do with what makes one organism different from another. These "boring" genes are part of the core genome of an organism; they might not be boring to a botanist who wants to know what makes a plant a plant, but they have little importance to what makes a cannabis plant at the genomic level. The core genome of plants in general can be estimated by comparing the genomes of a range of different plants and determining which genes occur in all of the genomes studied. Another thing that can be done when comparing genomes within a species is to sum all the different genes in the various individuals to describe the genomic distribution of the species. This result yields what is called the pangenome of a species (fig. 12.2). Pangenomics has become an important part of studying cultivated

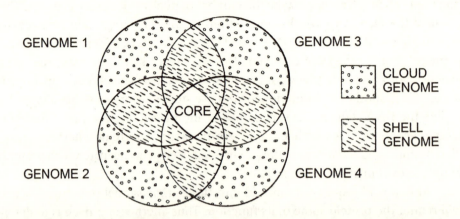

Figure 12.2. Venn diagram representation of the pangenome, core genome, shell genome, and cloud genome of a group of related organisms. The diagram shows the core genome in the middle of the diagram, which is simply the overlap of all the genomes in the comparison. The shell genome consists of any genes that are in two or more of the individuals in the study. The cloud genome is the collection of those genes that are present in only a single individual species. Here there are four cloud genomes, one for each of the individuals included in the comparison. The pangenome is the sum of all the genes in the diagram.

organisms, because the pangenome will indicate the full extent of genomic potential a species might have.

If one wants to understand what makes cannabis the way it is, then the nonshared parts of the genome will reveal where all the action might be. If some genes are missing in one plant genome, then this might correlate with differences in anatomy or function. Other categories of genes that are of interest are those that exist in only a single individual or are representative in a comparison (cloud genome), and those that are found in at least two of the individuals in the comparison (shell genome). Much of modern genomics so far has been accomplished using what are called reference genomes. These are based on the genome of a single or a few individuals and have provided a good starting point for modern genomic studies of a wide range of organisms. But a species such as cannabis has a lot of variation in the genomes of its members. The variation is what is important for a genomicist; it can be used to determine the genetic basis of traits of the cannabis plant and to infer certain aspects of its genealogical history.

A single reference genome simply does not do justice to the degree of variation that a species such as cannabis might have; a pangenome does. For instance, one of the principal goals in studying cannabis gene function is determining not just the presence of a gene or gene category but how many copies of the gene or category exist in the genomes of the many strains of cannabis. This category of data, called copy number variation or CNV, enables a cannabis researcher to determine if the number of THC genes in the genome of a cannabis strain has anything to do with the THC content of the strain. In addition, the position of small and large insertions in the genome affects how the cannabis genome works to produce the plant's many chemotypes and phenotypes. The order of genes on the chromosome is also important for understanding both function and the history of cannabis. This aspect of a genome is known as synteny; when two strains have the identical order of genes on a chromosome, they are said to be syntenic. The more synteny two organisms share, the more likely they are closely related.

By 2020 eleven high-quality and somewhat thoroughly annotated cannabis genomes had been sequenced. In addition, another two cannabis strains have had their transcriptomes sequenced. These latter sequences are obtained directly from the transcription products (the messenger ribonucleic acid, or mRNA, synthesized from the DNA of genes in the cells) of the genome. They represent which genes in the genome are expressed—in other words, which genes are active. These are listed in table 12.1, along with other statistics about the genomes of the strains involved. If the strains are variable enough, the eleven genomes can be scanned for single nucleotide polymorphisms (SNPs), which are the currency of modern genomics.

Single nucleotide polymorphisms are simple to understand; they are positions in the genes and intervening regions that differ in one or more reference individuals. But it is often tricky to determine where these SNPs are and whether they will be useful in yielding information relative to the genomes of individuals that are not part of the reference group. In addition to the eleven genomes mentioned above, Kevin J. McKernan and colleagues sequenced forty complete genomes of a wide range of cannabis cultivars to add to the pool of fully sequenced cannabis genomes. Of all of these cannabis genomes available in 2022, three in particular have been used to understand the genome dynamics of cannabis. Table 12.1 shows that these top three strains with the fewest "contigs" and the larger scaffolds are cs-10, Purple Kush, and Finola. Fewer scaffolds indicates that the assemblies are more complete and interpretable. If we liken the genomes in table 12.1 to 500-piece puzzles, the smaller scaffold numbers mean that there

Table 12.1. CANNABIS REFERENCE GENOMES AS OF 2022

Cultivar	Sex	Total sequence length	Scaffolds	No. of chromosomes	Coverage	Platform	No. of proteins
cs-10/CBDRx	f	876,147,649	221	10	100x	Oxford Nanopore	25,302
Purple Kush	f	891,964,663	6,653	10	79x	PacBio	30,074
Finola	m	1,009,674,739	2,362	10	98x	PacBio	37,689
Bubba Kush	m	512,174,223	18,355		72x	PacBio	
LA Confidential	f	595,358,288	311,039		50x	454	
Chemdog 91	f	285,932,793	175,088		50x		
Cannatonic	f	585,823,666	11,110		10x	PacBio	
Jamaican Lion (mp)	f	876,735,611			125x	PacBio Sequel	27,664
Jamaican Lion (fp)	m	1,009,156,132			125x	PacBio Sequel	31,591
Jamaican Lion	f	999,122,115				PacBio Sequel	
JLd	f	812,525,420	483	10	153x	PacBio Sequel	38,382

Note: The cultivar sequenced is listed first. The sex of the plant is listed next, with the total number of nucleotides sequenced in the third column. The fourth column lists the number of contiguous "scaffolds" that were obtained upon assembly. The lower the number of scaffolds, the higher the quality of the assembly (i.e., fewer gaps in the overall sequence). The fifth column lists the number of chromosomes that could be inferred from the sequence data, and the sixth column lists the x-coverage of the project. "Platform" refers to the kind of sequencing technology that was used. The last column lists the number of proteins inferred from the annotation process.

are fewer lost pieces and more of the different areas of the puzzle have been put together.

The process of finding where the valid SNPs are located in the genome for a species is called ascertainment; the SNPs are collected or ascertained given a set of rules that ensure that they are variable enough to be useful (and also not the result of errors in the sequencing process). Once researchers have pinpointed where the SNPs are for a species, it is much easier and cheaper to obtain genome level information for more individuals. This process of targeting specific variable regions of the genome is called genome resequencing. In addition, technology can use where the SNPs are known to exist followed by the sequencing of those regions only. This technology allows for more rapid, extensive, and cheaper genome sequencing for cannabis, with its thousands of strains. Similar methods have been used to sequence portions of genomes in other organisms that can produce recreational beverages, including grapes, yeast, rice, and barley, which each have thousands of cultivars.

It has been known for some time that cannabis has ten pairs of chromosomes. One of the first attempts at determining the number of chromosomes in cannabis occurred about a century ago by a Russian cytogeneticist, Lidiya Petrovna Breslawetz, who used microscopy to view and count the chromosomes. That number was an even twenty in a diploid cell, which meant cannabis had ten pairs of chromosomes. Another way to pin down the number of chromosomes in cannabis is to take its genome sequence and calculate it from there. Marijuana genomicists unfortunately have tried to order the chromosomes of the cannabis genome in a rather confusing way, as the different reference genomes have adopted different ways to number the chromosomes. Having the same numbering system is important when one tries to compare the chromosomes across cultivars.

As mentioned earlier, humans have a total of twenty-three pairs of chromosomes. These consist of a pair that can have two sex chromosomes that are called the X and Y chromosomes. The remaining twenty-two pairs of chromosomes (autosomal chromosomes) are numbered according to their size; the largest (chromosome 1) down to the smallest (chromosome 22). Cannabis also has one pair of sex chromosomes (called the X and the Y), and the leftover nine pairs of autosomes make up the cannabis genome at the level of chromosomes. To simplify matters, researchers have fallen back on the Finola numbering system because it truly reflects the size of the chromosomes in ascending order (table 12.2).

Once you have a highly contiguous genome assembly (meaning the jigsaw puzzle has only a few missing pieces), determining what genes are where (annotation) can proceed, and this is where most of the fun starts. Much of the interest in the cannabis genome results from curiosity about its so-called 1,001 molecules. These molecules code for the proteins involved in synthesizing most of the cannabinoids and terpenes found so far. But cannabinoids are only a small subset of the 1,001. The larger general class of molecules that are of interest are called phytochemicals, and genes for their synthesis make up a considerable part of the cannabis genome. Phytochemicals exhibit amazingly diverse makeup and shapes,

Table 12.2. THE TEN CHROMOSOMES OF THE FINOLA
STRAIN OF CANNABIS

Chromosome	Size (mb)	Genes
1	100,649,945	2,893
2	95,692,043	2,894
3	94,587,701	3,346
4	92,111,078	2,572
5	87,047,438	5,117
6	77,135,887	2,218
7	76,634,836	2,915
8	76,024,397	2,519
9	49,536,295	2,576
10	35,338,263	1,937

and are extremely versatile in their biological effects and physical attributes. They are an integral part of a plant's chemical personality. An important step in understanding how these phytochemicals are synthesized and work is to locate them within the tangly forest of the cannabis genome by using gene sequences.

The Inner Workings of the Cannabis Genome

If you have a basket of apples and oranges and you want to separate them, it is usually a simple task to sort them on the basis of color. But even if you are blind or color-blind, it still is possible to sort them based on other characteristics such as shape, stem characteristics, and so on. You might also use the texture of the fruits to determine which is an apple and which is an orange. Sorting out two kinds of things is easier the more different they are. Genome annotation resembles sorting apples and oranges in the genome, using our background notions of what this gene or that gene looks like.

But instead of just apples and oranges, there are hundreds of other fruits in the original baskets, and frustratingly different kinds of each fruit (with apples, for example, there are Empire, Gala, Fuji, and nearly 7,500 other cultivars). Such gene families are similar in sequence and often have similar functions. The synthases that eventually produce THCA and CBDA are examples of genes that have other gene family members, and so they pose small problems with the annotation process. But they aren't the only genes in the cannabis genome that are members of families, and this fact complicates annotation even more. At the heart of annotation is the recognition of similarity of sequences. If a gene has a threshold similarity to another gene at a particular cutoff (say 75% similarity versus 90% similarity), it can be considered in the same gene family, but if the cutoff values are too lax, the gene in question might be unique with respect to other genomes.

Tables 12.1 and 12.2 show that the annotation process for cannabis is relatively complete. The number of genes on each of the ten chromosomes ranges from over 5,000 to just under 2,000. But note that the gene numbers do not reflect chromosome size. For instance, the fifth largest chromosome (chromosome 5) has the most genes of any in the genome (over 5,000), and the largest chromosome (chromosome 1) has a rather average number of genes at a little under 3,000. This disparity between chromosome size and gene number is caused by many factors, mostly related to the amount of intervening sequence found between genes. Another factor might be that some chromosomes carry more repetitive DNA. Geneticist Rahul Pisupati and colleagues have examined the distribution of repeated elements in cannabis genomes and have shown that a wide array of these elements take up about 64 percent of the entire genome. The question is what the repeated elements are doing.

Jumping Genes

If you had the good fortune of visiting Cold Spring Harbor Laboratories on New York's Long Island in the middle to end of the last century, you might have come across a legend: the plant geneticist Barbara McClintock (fig. 12.3). In the middle of the last century, we didn't know much about how DNA was involved in genetic transmission, so most of the researchers of that period focused on rather simple genetic problems. Not McClintock. She was intrigued by corn and the myriads of colorful outcomes of the kernels that maize breeding could produce. She became close to maize in a highly personal way (her biography by Evelyn Fox Keller is titled *A Feeling for the Organism*). She also focused on a phenomenon in maize

Figure 12.3. A Swedish stamp honoring Barbara McClintock. Note the corn and the mottled appearance of some of the kernels. This is the trait that was exceedingly difficult to pinpoint, but she was able to do it.

called variegation or mottling, which occurs when the corn kernel has a consistent background but features the mottled appearance of a different color. It was a mysterious phenomenon, and McClintock set out to understand how and why this occurred.

She worked strenuous, long hours on the problem. When given the chance to explain the variegation of the kernels, she was met with skepticism and derision. One attendee at one of her presentations said, "I don't want to hear anything about what you're doing. It may be interesting, but I understand it's kind of madness," implying that she was insane. Another famous molecular biologist called her "just an old bag who'd been hanging out at Cold Spring Harbor for years." She endured unfortunate, disdainful, and sexist behavior from her male colleagues, but McClintock won a Nobel Prize for her work in 1983, getting the last laugh.

What was so innovative about her work? She was able to decipher how the mottling on her corn kernels was produced by using complex genetic crosses and ironclad reasoning to suggest that there must be some sort of moving genomic element inserting itself near the kernel color genes. These elements were disrupting the expression of the pigment genes to produce the mottling. It was known at that time that large-scale inversions of chromosomes move genes around a bit in organisms, but she ruled out large-scale chromosomal inversions by closely examining the chromosomes. She observed no big inversions, and the genetics of the mottling trait behaved quite differently than large inversions did. She then hypothesized that some sort of phenomenon could move along the chromosomes, because she was able to map the controlling factor for variegation and watch the factor move as she examined more of her maize crosses.

The genes controlling the variegation were "jumping" from one part of the chromosomal makeup to another. This was quite stunning for two reasons. First, she did all of this by manipulating crosses in small plots of maize, and she used a microscope to view the chromosomes. Second, she discovered a genetic mechanism that was so nonintuitive that her diligence made her the only possible person to understand it. Since her work, this "jumping" phenomenon and its impact on traits in organisms has become a basic tenet of genetical inheritance. She certainly stands alongside Gregor Mendel, Thomas Hunt Morgan (the discoverer of crossing-over in flies) and Hermann Joseph Muller (who worked out the basis of genetic mutation) as a giant on whose shoulders many modern biologists have stood.

McClintock's importance to our cannabis story goes back to the 64 percent of the genome that is repetitive. Mobile elements, transposable elements, or jumping genes exist in the genomes of some plants in copious numbers. There are often thousands to millions of copies of any given mobile element in any single plant genome. These mobile elements are disruptors; they frequently transpose into the coding region of genes themselves or in regions of the genome that control genes' on–off switches. These transposable elements fall into several categories that have different characteristics and ancestries. They paste and copy themselves, then excise, move, and repaste in a different part of the genome.

Long terminal repeats are a major category of transposable elements, and elements such as Ty1-copia and Ty3-gypsy (one named for its abundance and the

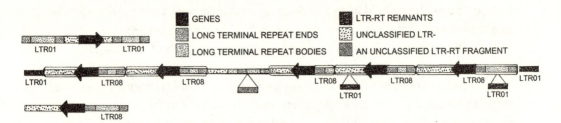

Figure 12.4. Genetic map of seven CBD synthase genes (black arrows; genes represented by black). The different kinds of long terminal repeats (LTR) in close proximity to the CBD synthase genes are indicated by the different shaded boxes. The black off bars indicate decaying LTRs. Adapted from Grassa et al. (2021).

other for its capacity to wander around the genome) make up most of the transposable elements in this category. There are also a large number of simple sequence repeats (SSRs) that constitute a reasonable part of the repetitive DNA in plant genomes and hence in the cannabis genome. Transposable elements are not only important for explaining why the genome of cannabis is as big as it is, but also they can influence the expression of any gene they might disrupt. Indeed, some important cannabis genes are distributed in the midst of long terminal repeat "junkyards," as shown in figure 12.4 where seven CBD synthase genes are embedded among dozens of long terminal repeats.

Transposable elements might be involved in how cannabis increases the concentration of the molecules that we find important for its use as a medicinal and recreational substance. Much of cannabis breeding and artificial selection have focused on getting more and more CBDA and THCA in different strains of the plant. To understand how to improve the cannabis phytochemical repertoire, we need to learn about the pathways that are responsible for the big two, CBDA and THCA. And there are other molecules in the 1,001 that are significant factors in odor, taste, volatility, growth parameters, and other variables. These are all traits that can be further understood by using the genome of the plant.

13

Putting the Cannabis
Genome to Work

Once you have a genome sequence, what kinds of questions can you answer? Although the idea that sequencing the genome is the equivalent of understanding everything about an organism is false, there is much you can discover from a sequenced genome. By examining the genome's structure, you can determine how its major features are arranged. These features include genes (the sequences that code for proteins), noncoding regions (sometimes referred to as the genome's "dark matter" because their functions are still not fully understood), and transposable elements (DNA from other genomes that has inserted itself into your organism's genome; viruses are one well-known example). Understanding the arrangement of these elements can actually reveal quite a lot about how genes function to create the organismal traits that interest us.

This process is called structural annotation: a researcher analyzes the genome sequences (or rather trains their computer to analyze the sequences) and uses the known properties of genes to identify where in the genome they are located. The genome is a long string of Gs, As, Ts, and Cs, but proteins are made of twenty different amino acids. How do we get from four nucleotides to twenty amino acids? One convenient aspect of DNA and how those four nucleotides (G,

A, T, and C) are translated into proteins is that a code using the nucleotides has evolved over billions of years of life and is the same or quite similar for almost all living organisms. The DNA somehow has to code for those twenty protein building blocks and to put them in the right places. What makes some of this easy is that DNA and proteins are both linear, and the positions on the DNA strand are correlated with the position of an amino acid in a protein. This means that the DNA that codes for the first amino acid of the protein will appear at the beginning of the DNA strand.

We know the code can't be one-to-one: it isn't possible for four nucleotides to directly code for twenty amino acids. You might think of some wild ways in which a single nucleotide code might work to code for a single amino acid in a protein, but they don't occur in nature. A two-nucleotide code is also unfeasible, because there are only sixteen possible two-nucleotide codes—AA, AT, AG, AC, GG, GT, GA, GC, TT, TA, TG, TC, CC, CT, CA, and CG—which can't cover the twenty amino acid protein alphabet. If you consider triplets such as AAA, AGA, and AGG coding for a single amino acid, then it is possible to code for twenty different amino acids—but now the problem would be that there are sixty-four three-nucleotide codes, but only twenty amino acids.

In fact, this was an easy problem for evolution to solve—instead of one triplet coding for only one amino acid, the "extra" triplets are used as different ways of coding for the same amino acid. So, for instance, the amino acid proline is coded for by CCC, CCG, CCT, and CCA. Any one of those triplets will produce exactly the same amino acid. The genetic code is therefore said to be redundant, because there are multiple DNA codes that will lead to the same protein building block.

This triplet-based genetic code is well characterized. We refer to each of the sixty-four triplets as a codon, because each codes for a specific product. For a few codons, the product is not an amino acid but rather an instruction. Since coding and noncoding DNA both consist of strings of As, Cs, Ts, and Gs, how does the genome "know" where to start treating codons as amino acid codes? There are codons in the genetic code that indicate where to start and stop. Every coding region will have the following structure: start codon . . . some number of amino acid codons . . . stop codon. These stretches are called open reading frames (ORFs) because they tell the cell where to "read" for instructions when building a protein. If your organism has had its genome sequenced, then searching for proteins is as simple as programming your computer to translate *in silico* all possible stretches of your organism's DNA into proteins and looking for long stretches of translated amino acids that are bracketed by start and stop codons. But there are six ways the DNA of a genome can be translated. Because the code deals with triplets, there are three points where the translation can start. And there are two strands of nucleotides in genomic DNA, so that makes another three points where the translation can start. To further complicate matters, the starting point can change from genomic region to genomic region. It is difficult to sort through these variables as a human, but a computer can do it in a snap.

Once you have identified all the ORFs, you try to match the amino acid sequence of the putative proteins to a reference database of known protein sequences. Most of the proteins in these databases have been studied in great detail, and their functions are well understood. If you find that your putative protein is highly similar to a reference protein, you can use that information to infer the function of your newly discovered protein. This process is called functional annotation, and it works because proteins with highly similar amino acid sequences tend to have very similar functions.

This reference-based functional annotation process has been used to identify many of the genes involved in the production of cannabinoids, lignins, flavonoids, and terpenes in the cannabis genome. Epigeneticist Igor Kovalchuk and colleagues generated a map (fig. 13.1) that summarizes what is known about the locations of these genes. Beyond these simple principles, however, things can easily get more complicated. For instance, what if one putative protein sequence is highly similar to not one but eight different reference proteins? Those eight reference proteins must have highly similar sequences, or else you would not have had matches to all of them in the first place. Based on the principle above—that similar proteins have similar functions—you can conclude that all eight of the reference proteins are in the same gene family and have similar functions. This may occur because proteins evolve through common ancestral proteins; often it occurs by duplication of the gene for the protein. As a result, you might find several proteins in a row (linked or in close physical proximity on the same chromosome) that have extremely similar sequences. This is indeed what happens in the cannabis genome, where several cannabinoid synthase genes (CBDAS, THCAS, and CBCAS) are found in close proximity to each other on cannabis chromosome 6.

It should be noted that while knowing the location of these genes is important, this information is essentially only an ingredients list (and a crude one at that) for use in the production of a cannabis plant. Variation in how these genes act to produce and sequester the compounds they code for is what gives different marijuana strains their particular characteristics, and this information can't be

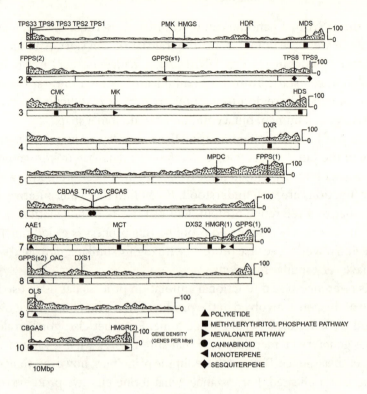

Figure 13.1. The Cannabis genome, adapted from Kovalchuk et al. (2020). Abbreviations: AAE1, hexanoyl-CoA synthase 1; CBCAS, cannabichromenic acid synthase; CBDAS, cannabidiolic acid synthase; CBGAS, cannabigerolic acid synthase; CMK, CDP-ME kinase; DXR, deoxyxylulose phosphate reductoisomerase; DXS, deoxyxylulose phosphate synthase; FPPS, farnesyl pyrophosphate synthase; GPPS, geranyl-pyrophosphate synthase; HDR, hydroxymethylbutenyl diphosphate reductase; HDS, hydroxymethylbutenyl diphosphate synthase; HMGR, hydroxymethylglutaryl-CoA reductase; HMGS, hydroxymethylglutaryl-CoA synthase; MCT, methylerythritol phosphate cytidylyltransferase; MDS, MECDP-synthase; MEP, methylerythritol 4-phosphate pathway; MVA, mevalonate pathway; MK, mevalonate kinase; MPDC, mevalonate diphosphate decarboxylase; OAC, olivetolic acid cyclase; OLS, olivetol synthase; PMK, phosphomevalonate kinase; THCAS, tetrahydrocannabinolic acid synthase; TPS, terpene synthase.

understood simply from knowing the genes' location. Staying with the cooking analogy, the ingredients are relatively meaningless until the chef decides how to prepare them. Likewise, the genes involved in producing cannabis compounds can be combined in myriad ways through changes in when, where, and what quantity they are expressed. The combined knowledge of which genes are involved and how they interact to produce cannabinoids (or any other important trait) is called the genetic architecture of the trait, which requires research methods beyond those we have so far discussed.

Before molecular biology started to become relevant to genetic analysis, even determining the location of the genes responsible for a trait was cumbersome. Researchers usually performed crosses between individuals with different versions of the trait (for example, high versus low sugar content in corn) and analyzed hundreds of individuals in each generation of a study. They needed to use organisms that bred quickly, to get as many generations as possible crammed within a reasonable amount of time. Researchers also needed markers that were dispersed on the chromosomes of the organism's genome. The crosses took advantage of a cellular process called recombination: chromosomes line up with each other in a cell, and while they are in close contact with each other, information can be swapped between them. This creates new combinations of genes on a chromosome, hence the term recombination.

Recombination between any two genes mostly depends on how close together they are on the chromosome. Next-door-neighbor genes are unlikely to recombine, because there is only a tiny region where a recombination event would separate them, while for genes located at opposite ends of a chromosome, any recombination event will separate them. By using visible markers to track recombination events, early geneticists could determine whether the gene for a specific trait was "close" to one of the markers. For instance, for a chromosome with three markers, if crosses show that the trait co-occurs with marker 1 we can conclude that the gene of interest is physically closer to marker 1 than to marker 2 or 3. We won't know the exact distance between the gene and marker 1, but with a more dense set of markers we can derive increasingly precise estimates of where the gene is located. The underlying theory is simple, but this approach only works if the number of individuals in the study is large. Unfortunately, undertaking crosses that produce hundreds or thousands of progeny is not possible with most organisms. But in cases where it is feasible (most famously in fruit flies, nematodes, and yeast), this approach has enabled geneticists to map the genes responsible for many traits with high precision. Barbara McClintock used these techniques to discover the transposable elements in maize. But even when successful, the physical mapping procedure is quite time- and labor-intensive; imagine studying hundreds of thousands of flies to determine the location of a single gene.

One obvious example of an organism where controlled crosses and huge progeny populations are not possible is human beings. The logistical and ethical issues are insurmountable in humans, so other methods had to be developed to map human genes by examining how traits are distributed within families. For instance, if over four generations of a family's history the only people who died from heart attacks were Aunt Bertha, Grandma, Great-Grandma, and Great-Great-Aunt Sally, and they all had red hair while all other relatives had brown hair, we can hypothesize that there is a gene for heart disease located near the gene for red hair (and that it is only expressed in females).

As modern genomics methods emerged, researchers interested in finding the genes that cause human disease began to envision a broader molecular approach to mapping genes. Their first approach was to try to replace visible markers (such as red eyes in fruit flies) with molecular markers. Molecular markers are "visualized" by genetic sequencing, rather than by inventorying an individual's appearance. It is generally much easier to find small genetic differences between organisms than to find visible differences, so molecular markers provide a much denser set of values—instead of three markers per chromosome, one can easily generate tens or even hundreds of molecular markers. This approach is called quantitative trait loci mapping, because it enables one to consider the contribution of many genes to a single trait. Traits are quantitative if they differ in degree, like height or weight. Qualitative traits are either/or, like having a widow's peak. As a general rule, quantitative traits involve the actions of many genes, while qualitative traits are controlled by one or a few genes. Quantitative trait loci (QTL) studies have two basic requirements: a variable, well-characterized trait, and a large set of variable molecular markers. Researchers begin by identifying strains that vary with respect to a particular trait; if one is interested in understanding why some plants express high levels of THCA, one would start with a single high THCA strain and a single low (or no) THCA strain. To determine if the available molecular markers are sufficient, researchers have to test whether there are enough genetic markers that differ between the high and low strain. For example, marker 1 can be identified as coming from the high versus the low strain. Ideally, the researcher will also have solid information about the genomic context of marker 1 (what chromosome it occurs on, what genes are located nearby, and so on). In other words, a relatively good genome map of the organism is required.

Once a researcher concludes that the traits are solid and that there is sufficient variation to carry out the QTL analysis, crosses are performed between the
two strains. Imagine a simplified case where strain 1 has a THCA concentration
that is high, and strain 2 has a THCA value that is low or zero. The first-generation
hybrids are called F1s, and they will each inherit half of their genome from strain
1 and half from strain 2. They will thus be heterozygous for both the trait (all will
have one high gene and one low gene, which will produce an intermediate trait
score of THCA = 5) and for the QTL markers (all have one copy of each strain 1
marker and one copy of each strain 2 marker). The F1s are then crossed in various
ways, and the second-generation offspring (F2s) are examined and scored for the
trait. In this simplified example, the laws of inheritance dictate that the F2 trait
scores will range from low to high. If we find a QTL marker that occurs in all of
the THCA = high F2s and none of the THCA = low F2s, we can conclude that the
marker is linked to the trait (that is, the gene for high THCA is located in close
proximity to the marker). This approach only works with organisms that can be
crossed to generate many different recombination events, because the goal is to
get the gene and the QTL marker very close to each other on a chromosome that
has no other genes from the high THCA strain. The QTL mapping approach has
enabled researchers to extend their work beyond standard model organisms such
as fruit flies and to start to answer questions about the genetic architecture of
traits of economic and medical importance.

In cannabis, QTL studies have been used to determine the genetic architecture of several traits. In hemp-type cannabis, plant breeding specialist Jordi Petit
and colleagues used QTL mapping to characterize the genetic architecture of vegetative time, flowering time, and sex determination, traits that are strongly associated with hemp fiber quality and seed production. They identified a set of QTL
markers associated with these traits, and by examining the genes located near the
quantitative trait loci they concluded sex determination was affected by genes
involved in light perception and genes that regulate flowering interact to control
flowering time, as well as genes regulating the balance of two classes of plant hormones (auxins and gibberellic acid). Also in hemp-like cannabis, crop geneticist
Patrick Woods and colleagues used a QTL approach to review variation in eight
agronomic traits (such as seed yield and stem biomass) and seventeen biochemical traits (such as terpene and cannabinoid concentrations). They found thirty-
four QTL markers associated with the agroeconomic traits, but these clustered
into just four regions; similarly, thirty-five QTL markers were associated with
variation in the biochemical traits, and these mostly mapped to two regions. Their

results suggest surprisingly simple genetic architecture for the traits they examined. In particular, a single candidate gene encoding olivetol synthase was found to be important in determining both agronomic and biochemical traits, indicating that this gene may play a key role in the genetic architecture of many traits of interest to cannabis breeders. However, another QTL study published in the same year had different conclusions. Bioinformatics researcher Christopher Grassa and colleagues performed a hemp-type × marijuana-type cross (as compared with hemp × hemp in the two previous studies) to determine the genetic architecture of THC : CBD ratios and total cannabinoid content. They found sixteen QTL markers associated with cannabinoid traits, but these had little overlap with the QTL markers identified by Woods et al.

These inconsistent QTL results probably reflect the fact that many of the traits cannabis breeders are most interested in are complex, meaning that the exact state of the trait depends on the interaction of many genes in the genome. Even exactly the same genes can have different effects depending on the genetic background in which a trait occurs, and as a result QTL studies can find effects that are valid in the cross that was performed for a particular study, but nonexistent in a different cross. This problem, along with the technical challenges of obtaining sufficiently large sample sizes, has led researchers to look for alternative approaches.

One of the most commonly used alternatives is the genome-wide association study (GWAS). In GWAS one simply surveys a large population for trait variation, sequences as many genomes as possible in the population, and examines the data to determine whether, for example, all the high TCHA plants share some genetic variants that are absent in the low THCA plants. The principle is similar to QTL mapping, but no crosses are required. In a species (like humans) where controlled crosses can't be performed, or in one where generating large sample sizes from controlled crosses is difficult, GWAS can enable researchers to begin to understand the genetic architecture of a trait. This approach underlies many of the recent discoveries of genes associated with an increased risk of heart disease, cancer, and other human ailments. Just as with QTL mapping, a GWAS is convincing only if the traits of interest have been meticulously defined and measured. This requirement for well-studied, carefully characterized traits is essential, no matter what technological approach is employed.

The need for highly reliable trait data for cannabis to investigate the genetic basis of the traits has both an upside and downside. The downside is that precise data on cannabis phenotypes have only recently become publicly available, in contrast to other crops such as corn (see chapter 12), where almost a hundred years of research have produced many strains of maize with precisely defined phenotypes, making the genetics of complex traits in corn a fairly tractable question. In cannabis such research has been held back by the long-standing criminalization of the growing, selling, and consumption of cannabis—no researchers in their right minds would risk incarceration for a significant biological study of cannabis. As a result, only a small group of people have studied cannabis traits—mostly those who grow, sell, or consume it. Much of this information on trait variation in cannabis is therefore anecdotal, subjective, and potentially unreliable.

There are numerous websites where the phenotypes of different cannabis cultivars are described in great detail. Evolutionary biologist Matthew Aardema and I wondered whether this anecdotal information could be successfully used as trait data for GWAS studies. We therefore collected trait data from a variety of marijuana-centric websites (including I Love Growing Marijuana [ILGM], CannaSOS, Leafly, Grow Marijuana, NCSM, Allbud, Wikileaf, and Cannabisinfo .com; see also For Further Reading for this chapter). The traits we examined included the CBDA and THCA levels in different strains, which account for some of cannabis's medicinal and psychoactive effects. As you might expect, the cannabis websites are heavily focused on this biochemical aspect of the plant, but not exclusively so—they also report on a strain's growth habits, its taste, its effect on the user's mood, and its usefulness in treating medical issues. For example, table 13.1 lists the information CannaSOS offers about the strain called Charlotte's Web.

By scouring the web for trait data such as these we were able to obtain 201 values for almost 300 individual strains of high THCA cannabis that also had genome sequence data available. Our question was: are these anecdotal trait data of high enough quality to be used in a GWAS study of the genetic architecture of trait variation across these 300 cannabis strains? As mentioned above, GWAS studies begin by surveying trait variation in a population; they then attempt to associate trait variation with genetic variation. If either type of variation is absent or unreliable, the study's conclusions will not be valid.

For example, suppose we are interested in the genetic architecture of lemony taste in cannabis. We will need trait data for as many strains as possible (perhaps lemoniness scored on a scale of 0 to 5), and the trait must be variable in the population. We will also need genetic data for these same strains, and not all genetic data are equally accurate or thorough, so this is another instance where avoiding "garbage in" is critical. Having collected these data for many different strains, we next divide the strains into "case" and "control" groups, just as in a

Table 13.1. TRAITS FROM THE CANNASOS WEBSITE FOR CHARLOTTE'S WEB STRAIN OF MARIJUANA

THC	3.72
CBD	11.72
Positive effects	Euphoria, uplift, focus, creative, energetic
Negative effects	Dry mouth, dry eyes, anxious, dizzy, headache
Flavors	Pine, earth, lemon
Symptom alleviation	Stress, pain, fatigue, muscle spasms
Flowering time	60–70 days
Height	72 inches
Yield	1.48 oz/ft squared
Growth difficulty	Moderate

human study where cases might be people with heart disease and controls would be people without heart disease. In our hypothetical study of lemoniness, the case group = strains with a lemony taste; control group = nonlemony strains. Having grouped our trait data in this way, we next look for genomic variation that distinguishes cases from controls.

As illustrated in figure 13.2, the genomic variants that are most commonly assessed are single nucleotide polymorphisms (SNPs), which are single base pair differences at a specific genomic location (e.g., having an A versus a T at position 500 on chromosome 6). Although it may seem incredible that such a small difference would have an effect, this has proven to be the case in many studies, including the one that identified olivetol synthase as a candidate gene for several cannabis traits. Once we find SNPs that distinguish our lemony versus nonlemony strains, how can we know whether they have any effect on lemoniness? After all, we expect to find a lot of variation between strains, and not all of it relates to lemoniness.

Deciding whether or not a SNP is "significant" in a GWAS study is not as simple as one might think. A SNP's significance score is basically a statement about its frequency in the case versus control groups. If all the cases have an A at position 500 on chromosome 6 and all of the controls have a T, then the SNP is clearly

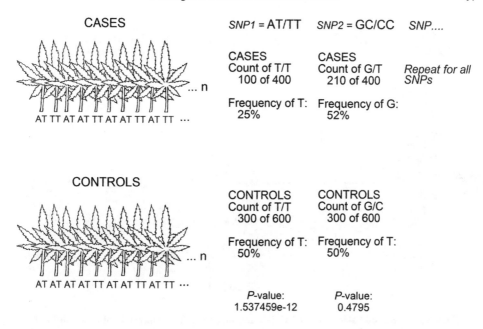

CASES

SNP1 = AT/TT   SNP2 = GC/CC   SNP....

CASES Count of T/T 100 of 400	CASES Count of G/T 210 of 400	Repeat for all SNPs
Frequency of T: 25%	Frequency of G: 52%	

AT TT AT AT TT AT AT TT AT TT ··· ... n

CONTROLS

CONTROLS Count of T/T 300 of 600	CONTROLS Count of G/C 300 of 600	
Frequency of T: 50%	Frequency of T: 50%	

AT AT AT AT TT AT AT TT AT TT ··· ... n

P-value: 1.537459e-12	P-value: 0.4795

Figure 13.2. Example of a GWAS for cannabis. SNP1 is an AT polymorphism in the first position of the sequence. In this example 1,000 cannabis strains were examined; 400 of the strains have a lemony taste, and 600 do not. The statistic calculated in this case is a chi-square P-value. In a typical study there are thousands of SNPs that need to be examined. The larger the number of cases and controls, the more precise the study will be.

significant. But anything less straightforward requires statistics: is the observed frequency different than what we would expect by chance alone? This is what P-values try to quantify: if $P = 0.9$, the observed distribution matches what we would expect if the SNP state is random with respect to the trait; if $P = 0.00000001$ (10^{-8}), the SNP state is unlikely to be random with respect to the trait.

In the example shown in fig. 13.2, SNP1 has a P-value of $1.537459e^{-12}$, and SNP2 has a P-value of 0.4795. While only two SNPs are shown here, most GWAS studies involve thousands of SNPs, and a separate P-value must be calculated for each of the SNPs. Depending on the study, as many as 130,000 known SNPs can be tested for whether they are associated with a particular cannabis trait. Once P-values have been calculated for all the SNPs, the SNPs can be visualized in a Manhattan plot (see fig. 13.3), which shows the genome locations and P-values for association with the trait. Each point in the graph represents the P-value of a single SNP, and the significance level cutoff is shown by a dotted line. In this

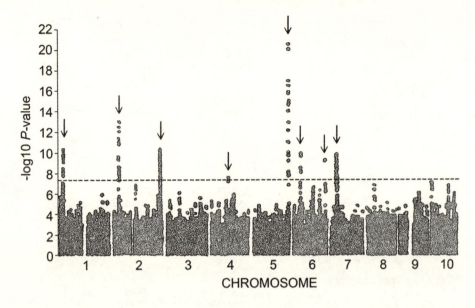

Figure 13.3. Made-up "Manhattan plot." Ten chromosomes and thousands of SNPs are represented in the diagram. The dotted line is the e10⁻⁸ significance cutoff. The points above the dotted line are SNPs that show significant association with the trait in question. The eight arrows
indicate potential genomic regions that could be associated with the trait.

made-up example there are eight regions of the genome where significant associations of the trait ("lemony") with SNPs in those regions can be inferred. Because
the positions in the genome of all the SNPs used in the study are known, a researcher can now go back to the genome map and discover which genes or genomic regions are close to the significant SNPs. These genes or genome regions
then become candidates for further study.

   Several studies have used GWAS for exploring the cannabis genome in hopes
of identifying candidate genes for the genetic architecture of important traits.
The most critical aspect of these studies is the careful selection (or generation) of
trait data. There are several possible routes. First, one can use trait data from the
literature or on the web, which is what Matthew Aardema and I tried in our study
using web-based trait data. Alternatively, one could insist on lab-based data from
specialized bioassays designed to precisely measure each trait. Obtaining SNP
data for marijuana strains is fairly easy, because several commercial outfits have
invested considerable money and time into sequencing the various strains that
they produce or sell. These SNP data are simple to download and organize. The

biggest issue is not in finding the genomic data of a strain, but rather determining exactly what strain the DNA sequences represent. The same strain may be called by many different names from one company or one grower to another, and if it isn't clear what strain the genetic data are from, the data are somewhat useless.

Our study on web-based trait data was sobering. Although there is a lot of data on these web sites, very few of the 200 or so traits we examined showed significant associations with SNPs in the genomes of the nearly 300 strains we checked. Most of the significant traits were for concentration of THCA and CBDA in the various strains. Part of the problem with traits other than the chemical ones is sample size (there are only a few strains typed as lemony), but most of the problem lies in the subjectivity or inaccuracy of the web-based trait scores that we used.

This result suggests that much more attention needs to be focused on characterizing interesting traits accurately and reproducibly. On the other hand, the study gave us hope that at least for biochemical traits, GWAS studies might be able to determine the underlying genetic architecture for those traits. In other words, if even the wobbly web-based biochemical trait data were able to reveal some SNP associations, once better trait data are generated the results are likely to be even stronger.

Two kinds of GWAS studies have been the focus of cannabis researchers who use the technique. Some researchers use the SNPs as markers to identify strains and then examine how those strains may have evolved over thousands of years. I call this "genomics as history." The second kind of study, just discussed above, can be called "genomics as architecture." Until we understand the genetic architecture of cannabis's most useful traits, we will be unable to precisely control what effect cannabis has on users. If one wants to use cannabis to control nausea in chemotherapy patients, for example, it is important to be able to provide that effect safely and predictably. Once we know the genetic architecture of such a trait, we are closer to being able to control and manipulate it through breeding or engineering.

Genomics as Architecture

One kind of study that attempts to associate traits with SNPs has been used to examine traits such as flowering time, sex determination, and fiber quality. I've already mentioned one of Jordi Petit and colleagues' studies on flowering time and sex determination. Their GWAS results indicate that there are six candidate genes for flowering time and two candidates for sex determination. This study was notable because the researchers completed a detailed analysis of flowering time, using plants grown in different environments and detailed phenotypic descriptions of the plants. The same group also examined fiber quality using GWAS approaches. While not a trait that recreational and medicinal cannabis growers might be interested in, fiber quality is an important trait in the hemp-as-fiber trade. Petit and colleagues were able to identify sixteen significant SNP associations with fiber quality traits such as genes involved in glucose, glucuronic acid, mannose, xylose, lignin concentrations, and bast fiber content. These sixteen SNPs lie in twelve candidate genes for fiber quality. Most of the candidate genes are involved in structures in the cell wall of cannabis fibers, and two-thirds are involved in lignin synthesis. Other genes involved in the architecture of fiber quality are related to monosaccharide or polysaccharide synthesis.

What about GWAS studies for the traits that make cannabis a desired recreational and medicinal substance? What about the architecture of cannabis traits that get us high? One way to approach these questions is to obtain precise phenotypic values for THCA and CBDA concentrations in cannabis strains with genome sequences available. This straightforward approach nevertheless is complicated by many issues, the most dominant being the strain name problem mentioned previously. If we have biochemistry scores for Polly Purple Kush, and genetic data for Pollyanna Purple Kush, we might want to combine them. But how do we know that they are the same strain? On the trait data side, there is a wealth of precise THCA and CBDA concentration data available for over a thousand strains. These data exist thanks to the strict rules around legalized marijuana in some states. For example, Washington State's rules for marijuana specify that "regardless of analytical equipment or methodology, certified labs must accurately measure and report the acidic (THCA and CBDA) and neutral (THC and CBD) forms of the cannabinoids." In a paper of immense value for cannabis GWAS researchers, Nick Jikomes and Michael Zoorob compiled an extensive collection of THCA and CBDA measurements across Washington State. While they found significant systematic variation in the values reported for the various strains, these concentration data might well be useful as trait data for GWAS studies investigating the genetic architecture behind the potency of strains.

Another approach is to use what is called an extreme phenotype GWAS, or XP-GWAS. This method examines only those individuals with an extreme phenotype, which in the case of cannabis could be very high versus very low THCA or CBDA concentration. Any intermediate strains are removed from the study, and then GWAS is performed as usual. This approach was used by plant geneticist Matthew T. Welling and colleagues in examining a strain's chemotype. Chemotypes, also called chemovars, rely on measurement of the THCA and CBDA content of a strain, using the ratio between these two as the defining trait. Three chemotypes are commonly recognized:

- ChemoType I—THC-predominant
- ChemoType II—balance of THC and CBD
- ChemoType III—CBD-predominant

Other criteria exist that define additional chemotypes such as these:

- Chemotype IV—a type III, rich in cannabigerol (CBG)
- Chemotype V—refers to cannabis plants that produce little to no cannabinoid content

Welling and colleagues used the XP-GWAS approach to obtain some interesting results concerning chemotype. The first and perhaps most significant is that the CBDA synthase gene was detected as significantly associated with chemotype—but so were several other loci. One of the more intriguing associations was with a gene called ACP (acyl carrier protein) that is involved in the polyketide pathway for synthesis of CBDA and THCAS. The results from this study suggest that XP-GWAS might be a valid technique to tease apart the genetic architecture of potency. Another inference that can be made is that while the approach did identify CBDA synthase genes as associated with genome SNP variation, it also revealed association of chemotype with a genetic element upstream of CBDA synthase, meaning that control of cannabinoids might involve a more complex genetic architecture than just the synthase genes at the end of the synthesis pathway.

All of this genome-level analysis is quite promising. Perhaps the genetic architecture of taste or chemotype will be worked out for cannabis, because once the architecture is known, the ability to manipulate it is just around the corner. There remains a caveat, though: better trait data are required for all of this to work. Kovalchuk and colleagues point out that part of the reason for the paltry amount of trait information in cannabis are the legal issues that prohibited academic study of the plant for several decades. This prohibition discouraged the collection of properly curated and archived cannabis material. Kovalchuk and colleagues summarize what is needed for future cannabis research: "Assembly of a global data set comprising various types of cannabis materials (e.g., undomesticated, naturalized, and landraces) as well as a comprehensive sampling of all putative infrageneric subgroups should be prioritized in future evolutionary studies, agricultural improvements, and medicinal applications."

Genomics as History

My colleague Ian Tattersall and I have taught a history of wine course at the University of Gastronomic Sciences in Pollenzo, Italy, every year since 2018. Tattersall is a wonderful storyteller, and his lectures on the ancient and not-so-ancient history of wine are especially appreciated by the students. My role in the course is to teach them about the genomics of grapes, which can be a tough sell for future chefs, sommeliers, and foodie entrepreneurs. What does genomics have to do with history? they ask. Everything! I tell them. The deep and not-so-deep history of the various grape strains and yeast strains that are used to make wine is written in the DNA of these organisms. Grape and yeast genomics underlie the convoluted story of wine making, peppered with root-stock fusions, hybridization, migration, artificial selection by breeders, and randomness. I also tell them that as long as you ask the right questions, it is hard to be fooled by the information embedded in DNA. So it is with cannabis history and genomics. But cannabis history is a bit more difficult to trace than that of grapes or yeast, because the breeding, cultivation, and transportation of cannabis have been necessarily unpublicized criminal acts until recently. Nonetheless, some progress has been made in the last decade using genomics to understand the genetic history of cannabis.

There are three basic ways in which geneticists have explored the history of lineages within a species. Many of these were developed to study the proliferating human genome information coming from academic and commercial labs. The first concerns evolutionary trees. Real trees have leaves, and so do phylogenetic trees. These leaves are often known as terminals or taxonomic units; in the case of cannabis, the leaves of any phylogenetic tree would be the different strains of cannabis. Real trees have branches and a trunk, and so do phylogenetic trees. The terminals at the end of a bifurcating branch are said to be each other's closest relatives. The point on the branch where there is a split is called a common ancestor of the two terminals. The trunk, simply a place where branches can spin off, contains common ancestor after common ancestor. (We should not confuse common ancestors with fossils, though. While a fossil that came before the two terminals on a branch could be a common ancestor, it most likely would not be found. Fossils sit on their own branches as leaves that have died out.)

The second tool is a model-based clustering method called *STRUCTURE*. This approach takes the sequences of all individuals and computes the probability that an individual is a genetic member of the potential populations being studied. Suppose you have some cannabis individuals for which you also have chemotype data. You want to investigate whether these individuals fall into two populations, recapitulating chemotype 1 and chemotype 2. The important part of the analysis is how many populations you think the individuals should be separated into. This is a parameter in the *STRUCTURE* analysis called *K,* which can theoretically be set to any integer value. Using $K = 2$, the probability of being a member of the first population and the probability of being a member of the second population are computed using a complex algorithm. These probabilities for each individual are then charted using a bar graph (fig. 13.4). If there are three or more populations, the researcher can reset and repeat the analysis with different *K* values. A statistical test would then be performed to determine which *K* is the best of those tested. The result is highly visual and useful in developing a good idea of what the genetic structure of a group of individuals might be.

The third method is called principal components analysis (PCA). This approach is a data reduction method, a particularly useful method for such complex datasets as genome sequences. DNA sequence datasets feature considerable variation. Indeed, without the variation such studies would be rather silly. The variation can be broken down into components, each of which "explains" a particular fraction of the variance in the dataset. For instance, a dataset comprising several individuals from two different populations can be analyzed with PCA. The analysis might show five components of variance, and these can be ranked 1 through 5. If one takes the variance measures from the first two components (perhaps the two best) for each individual and graphs them in two dimensions, a diagram such as the one on the left in figure 13.5 might ensue. Such a PCA result would indicate close genetic relatedness of the individuals. On the other hand, the points might be randomly distributed across the graph (as on the right, fig. 13.5). The tighter the distribution of the points, the more related those points are to each other.

The three techniques described here are the basic first-pass analyses that

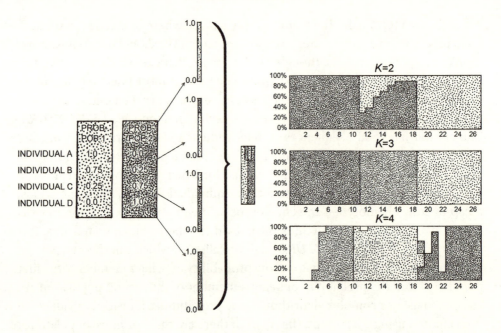

Figure 13.4. *STRUCTURE* analysis. The left panel shows how a structure diagram is displayed. The right panel shows a case in which twenty-seven individuals were analyzed for K = 2, 3, and 4.

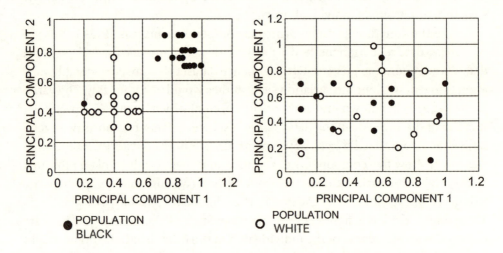

Figure 13.5. Two principal components analysis (PCA) graphs showing the first two components in each. The PCA on the left shows a distinct clustering of the individuals from the black circle and open circle populations. The PCA on the right shows a random distribution of the individuals from the black and white populations.

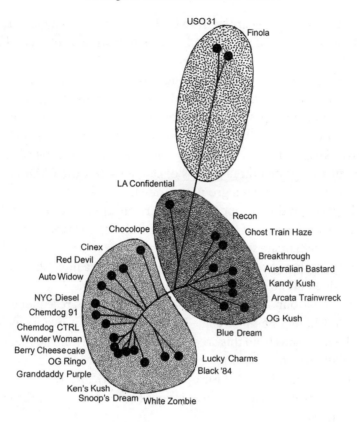

Figure 13.6. Three clusters of Canadian commercial strains.

geneticists use on any genome-level dataset. There are more complex and differently focused analyses that genomicists can perform with respect to a dataset. And more will be developed, depending on the needs of researchers in the future.

In 2015 biodiversity researcher Philippe Henry used the twenty-eight cannabis strains that the Canadian cannabis market had commercialized most frequently to review the history of marijuana chemotypes. He was able to detect three major clusters of strains, shown in figure 13.6. The first and most distinct cluster was the Finola-uso31 pair. These strains are colloquially (and ambiguously) known as Ruderalis, but are also well-known hemp fiber strains (i.e., strains with <0.3 THCA content). The other two clusters are medicinal and recreational strains. Henry did not examine Chinese hemp strains, but he implies that if he had there would probably be a fourth cluster comprising Chinese hemp.

But guess what? Six years later, two Chinese groups of genetic researchers addressed the dynamics of how the Chinese hemp strains are related to others. One group led by Jiangjiang Zhang used a "quick and dirty" genomic approach to examine the Chinese hemp strains along with European and American strains. Their approach, while scanning whole genomes, did not use the genome sequencing and resequencing approaches of other groups. Instead they used simple sequence repeats (SSRs), also known as satellites. These markers can be detected in a faster and less expensive way than whole genome sequencing or even resequencing. The results suggested three groups of cannabis, but the Chinese strains did not differentiate into their own group separate from other geographic areas. The analyses showed that the European and American strains were genetically interspersed among the clusters of Chinese strains. This suggests that geography is not the best way to demarcate strains of marijuana; there must be some other way to cluster the strains than geography. It also appears that the non-Chinese strains are deeply embedded in the clusters that are mostly Chinese.

The second group, led by Xuan Chen, used genomic resequencing to trace the origins of Chinese hemp strains. Their results suggest five major groups or clusters of hemp strains from this area. They unfortunately did not include strains other than Chinese ones, but their results clearly reflect the clustering they hypothesized. If they had included European and American strains, it is likely they would have observed the non-Chinese strains embedded among the five clusters of Chinese strains.

In an effort to further clarify matters, several other groups have recently used genomic sequencing and resequencing approaches to examine the ancient history of cannabis. One group led by Patrick Woods looked at over seventy strains from across the globe. They shied away from phylogenetic trees to do their analyses and instead used techniques that enabled them to examine the structure within the populations. Their results show further that the strains in use today originated in Asia. They considered wild American strains as escapees or feral varieties and compared them to domesticated strains. Using this approach, they were also able to show that the domesticated strains showed strong selections for certain traits related to increased THCA content.

Geneticist Dan Jin and colleagues examined almost 200,000 cannabis SNPs in twenty-three strains to review relatedness of the strains and whether chemotypes could be correlated to groups of strains. While they did not discuss specific associations, they pointed out that more than one-third of the significant SNPs lie on chromosome 6, which harbors the CBDA and THCA synthase genes. The other two-thirds of the significant SNPs were scattered over the other nine chromosomes of the cannabis genome. Another result they reported was that the strains tended to cluster into chemotype-defined groups. They concluded that while CBDA and THCA synthase genes might be involved in distinguishing the strains from each other, many other genes scattered throughout the genome were also involved.

Another group led by plant genomics researcher Anna Halpin-McCormick combined sequence data available from public databases for as many individual strains as possible. In total they analyzed over 2,500 different strains that had sequence data. Their results suggested that more structure could be detected in the genealogy of the strains when they used phylogenetic trees to treat these data. Depending on the way the analyses were designed, they obtained nine or ten clusters (also called clades), some of which held the majority of the strains examined. A caveat is necessary here, though. Branching phylogenetic trees are meant to organize differentiated units. Marijuana strains are hardly differentiated into separate entities, owing to the rampant hanky-panky through hybridization. This means that the underlying data are probably not appropriate for tree-building approaches. The principal utility of Halpin-McCormick and colleagues' study is that it demonstrated, with the largest dataset yet, how deep and far-reaching the hybridization among cannabis strains extends.

Perhaps the most important of the studies investigating marijuana population structure appeared in 2021 from phylogeneticist Guangpeng Ren and colleagues. They examined genome sequences of 110 different strains from all over the globe, using the three analytical approaches described above (fig. 13.7). Their results indicated that the strains clustered into four groups. So we now have three, four, five, nine, or ten clusters hypothesized by the different research groups discussed here. Ren and colleagues' study is particularly interesting because it indicates a logical way to reconcile the results from the different studies. Their four groups of strains consist of:

- A basic or basal cannabis cluster
- A hemp-type cluster
- A drug-type cluster
- A feral drug-type cluster

While the strains clustered into four groups in the tree analysis and the PCA analysis, the *STRUCTURE* analysis indicated a high degree of admixture or inter-

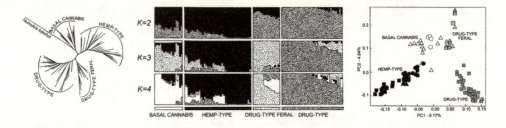

Figure 13.7. Three ways to analyze cannabis genome data. The left diagram shows a tree-based way to cluster the 110 strains analyzed in Guangpeng Ren et al. (2021). The middle diagram shows an analysis of the same data using the *STRUCTURE* approach for three *K* values (2, 3, and 4). The right diagram shows clustering of the 110 samples using the PCA approach.

breeding among the four groups (as indicated by the bleeding together of the different shadings in fig. 13.7). This way of organizing cannabis genome data makes a lot of sense, even though it does not follow geography. After all, the plant has been moved around continually, obliterating most if not all of its geographical structuring. In addition, it has been extensively hybridized and selected for certain traits by humans over the past 10,000 years—intensely so for the last century. It seems fitting that the organizing principals for analysis would be driven by human activities.

Genomics as the Future of Cannabis

Researchers' capacity to genetically modify plants (plant genetic engineering) will no doubt affect the future of cannabis. While the promise of genetic engineering is sometimes overblown, today's genome technology demonstrates a huge potential for modification of cannabis crops and enhancement of cannabis as a medicinal and recreational product. Of course humans have been genetically engineering cannabis for several millennia, and the THCA content of many marijuana strains has increased greatly in the last two decades. This increase has demonstrated that genetic alterations of cannabis can be quite rapid. Hundreds of distinct cultivars of the plant have also been established.

Because other agricultural plants have been altered genetically using biochemical approaches, it is not too outrageous to expect that modern molecular methods can also produce the genetic variations needed for further rapid and effective change in cannabis characteristics. At some point, though, genetic engineering using molecular techniques becomes an ethical and regulatory problem, and great care needs to be exercised in the future of cannabis cultivation.

There are several important requirements for effective genetic engineering of cannabis. First, the traits that are targets for engineering should be well understood and genetically accessible. A trait that is controlled by a thousand genes might not be a good target, given its complexity and the need to genetically modify the plant at these thousand loci. Creating the genetic variability for a trait with that extent of involvement of different loci is a huge task science has yet to conquer. By "well understood," I mean that we need to know the kinds of traits and genes that will be altered within the context of larger ecological and evolutionary systems. Many attempts to introduce altered organisms into complex ecosystems have been disastrous for those ecologies.

The reason this chapter and chapter 12 have focused on using genomics to map genes involved in cannabis traits is that once such genetic elements are found, their locations and sequences can be manipulated to create genetic variations that might lead to a cannabis plant with very precise genetic modifications. Once one has the gene or genes involved in a trait, technologies can be used to precisely alter those genes. But one must also understand the function of the gene product in order to know what to modify, and this task is not as easy as it might sound.

One technology used to make these precise alterations is called CRISPR-Cas9 (hereafter *crispr*). This technology takes advantage of a known gene sequence in the target organism, the known function of the gene product, and knowledge about the complexity of a trait to target specific single bases in a gene sequence. *Crispr* technology was developed in the early 2000s by biochemists Emmanuelle Charpentier and Jennifer Doudna. It uses a biochemical anomaly that certain species of bacteria have evolved to rid themselves of unwanted foreign DNA (e.g., from viruses). The enzymes involved in the *crispr* toolbox can be purified and used in test tubes. The researcher uses a guide sequence with a base that is a target for change. The *crispr* system then alters the target base and enables normal cell processes to repair the edited DNA. All this is done in a test tube, using the enzymes in the *crispr* toolbox. Once the desired construct has been created, it can then be inserted into the germ line of a target plant by a process called transformation. The past three decades have seen substantial advances in transforming

plant germ lines using a wide array of techniques, so getting the altered gene sequence into plant genomes and cells is entirely feasible.

Plants with the altered gene sequence are then grown and examined for the
desired trait changes. Researchers have had success using this technology to en-
gineer various plant traits, most of them in plants of agricultural importance such
as maize. Grapes are another example, where improvements in tartaric acid synthesis (a major acidic component of grapes), powdery mildew resistance, albino
phenotype, *Botrytis cinerea* resistance, and grapevine red blotch virus (GRBV)
resistance have all been engineered using the *crispr* toolbox.

Because cannabis is a newcomer to the legal agricultural lineup of crops, advances here are quite recent and focused on honing the engineering techniques so
that efficient transformation of cannabis plant cells goes smoothly. As discussed
in chapter 5, the sex life of cannabis also can complicate the breeding of a plant
that has been genetically engineered. For this reason, researchers are working to
establish doubled haploid (DH) technology for cannabis that will enable plant
breeders to rapidly create plant lines with homogenized genomes. The technique
requires a researcher to make a haploid plant (a plant with only one set of chromosomes, and hence only one parent) that can then propagate itself. This technology is essential for the true breeding of a novel genetically engineered cannabis plant. Since making a haploid plant is controlled by genes, some of the most
recent work on engineering plants has been focused on finding and engineering
these genes in cannabis.

These technologies are focused on engineering the cannabis plant itself to
manipulate traits that will maximize the plant's most desirable products. Plant
breeding methods are employed to harvest the results of the genetic engineering.
But other methods exist that are focused on artificial production of some of the
protein products of cannabis, most notably THCA and CBDA. Simply put: if one
is only after THCA, then the stems, seeds, leaves, and stalks are wasted biomass.
Several researchers and enterprising companies have turned to engineering microorganisms to produce the desired products of cannabis via bioreactor systems.
In these systems, researchers insert the genes for economically important cannabis proteins into bacteria, yeast, and algae and then let these microorganisms do
the synthesis work. Harvesting the end products (THCA and CBDA) then becomes simple biochemistry and avoids any cultivation of cannabis plants.

But think back to chapters 9 and 10, which discussed the biosynthetic pathways for plant cannabinoid synthesis. These aren't simple pathways. If an efficient
bioreactor is desired, the many steps in the synthesis of the cannabinoids all need

to be a part of the process. Nevertheless, considerable success has been attained in genetically engineering THCA and CBDA bioreactors for commercial use. One sign of its potential future growth is the number of patents submitted for accomplishing this method of CBDA and THCA production.

14

Modern Medicinal Cannabis

I suffer occasional bouts of gout. It is an arthritic autoimmune disorder brought on by the collection of urate crystals in some of your joints. It usually hits one of your big toes. It is intensely painful, almost to the point that you might want to saw off your big toe. Men are more susceptible to the disorder, first afflicting most of its victims in their thirties to fifties. Diets rich in high-caloric, fatty foods are usually a cause of onset, along with overconsumption of meats and shellfish. Imbibing too much beer or drinks with fructose may trigger it too. Other factors such as weight, diabetes, kidney disorders, and high blood pressure can trigger a flare-up. Gout can be slowed down by allopurinol, a compound that breaks down uric acid; colchicine, a compound that ameliorates inflammation; and corticosteroids such as prednisone, which also curtail inflammation. All these have the possibility of problematic side effects, but they are federally approved remedies.

I had read anecdotes about CBD and THC being effective at remedying the symptoms of gout. Apparently CBD offers some relief. If a little THC is added to the mix, then sleep becomes easier during a gout attack. (Usually gout is a real sleep killer, because even the contact of sheets with the surface of your toes produces a severe burning pain.) I contemplated trying this remedy for gout bouts, but one thing held me back: CBD and THC are not FDA-approved remedies for gout (nor for many other ailments). I could have done an "experiment of one" and tried CBD, but that is not recommended when medicine is involved. And if I felt better, I wouldn't be able to tell whether it was due to the CBD/THC or some other factor that I wasn't accounting for. The temptation to use CBD as a gout remedy, however, stimulated me to examine more deeply how cannabis medicinals are developed, tested, and approved.

A Sample Size of One

People have used cannabis for thousands of years as a remedy for one or another malady. The medicinal applications in East and South Asia are especially well known and wide-reaching. Many ancient medical texts from China outline the use of cannabis as a treatment for several maladies. India has also been a longtime source of anecdotal information about the medical uses of cannabis. It is this Indian knowledge that has made the biggest impact on how westerners think about cannabis as a medicinal, much of it thanks to one person—William Brooke (WB) O'Shaughnessy. He was an assistant surgeon in the East India Company, stationed in the city of Kolkata (Calcutta). Celebrated as the person who brought telegraph circuits to Kolkata at the Botanical Gardens in Shibpur, he was also a forensic poison expert and had stopped a cholera outbreak in Kolkata. Oddly, the original reason he went to India was because he could not pass his medical exams in England.

O'Shaughnessy published the first clinically based paper on the effects of medical marijuana in 1839, marking the birth of modern medical cannabis research. This paper simply described the many ways that he had observed people in Kolkata preparing and using cannabis as a medical agent. In some cases, he described his own administration of cannabis extracts to patients with various medical anomalies. He used anecdotal evidence to convey the range of medical effects of cannabis, thereby starting a long history of medical trials to assess the efficacy and safety of drugs and treatments for humans.

One report in his paper featured many of the aspects of a modern clinical trial. It concerned a convulsive forty-day-old female infant. The convulsions suffered by this poor, emaciated child were extreme and unrelenting, to the point that she could not sleep or eat. After using every possible remedy he had at his disposal (warm baths, doses of calomel, leeches, opium, and chalk), O'Shaughnessy decided that an alternate course of action was needed. He convinced the parents that hemp administration might be a way to treat their daughter, and they consented. A "single drop of spiritous tincture, equal to one-twentieth part of a grain in weight was placed on the child's tongue at 10 pm." When no immediate effect occurred, he added two more drops. She quickly fell to sleep and slept for what I calculate from O'Shaughnessy's reporting to be over twelve hours. Upon waking, she began screaming for food, was fed, and fell back to sleep again. The dosage set, he continued the application of hemp for the next four days, upon which all convulsions ceased, the child behaved normally ("the pulse, countenance and skin perfectly natural"), and she began to gain weight.

But a day after the child had been released to her parents, convulsions returned. O'Shaughnessy decided to up the dose once more (five drops every hour), but the convulsions only continued to occur. He then decided to examine the hemp medicine the child's servants were using to dose her. It turned out that instead of stoppering the medical vial with the rubber cap he had provided, the servants were using tissue paper. This caused evaporation and drying of the hemp extract, leaving water or a very diluted extract in the bottom of the vial. "The infant in fact had been taking drops of water during the previous day," O'Shaughnessy wrote. He mixed a new hemp preparation, administered it, and the child recovered as the dosage was adjusted. O'Shaughnessy pointed out that her treatment had become "manifestly a struggle between the disease and the remedy," as administering large amounts of hemp just put the baby to sleep. Indeed, some of the doses administered were huge, and according to O'Shaughnessy would induce effects as "profound as a trance in two men laboring under rheumatism." Once a proper dosage was found, the child recovered completely and "regained her natural plump and happy appearance."

This original "sample size of one" clinical trial had everything a clinical trial should have, although not in the best order. It involved a distinctive medical problem that could be quantified easily (convulsions), close observation of attempts to determine the right dose of a treatment, a placebo treatment (although not intended), and more dose adjustment to refine what was safe. The unintended placebo was serendipitously administered as a result of leaving the medicine vial capped with tissue paper.

While the origin of clinical trials and a scientific approach to understanding marijuana use in medicine goes back to O'Shaughnessy, several other historical events affected how cannabis was treated as a medicinal and how its use was eventually legalized. It is difficult to provide a cogent summary of the legal history of cannabis, because it encompasses many laws but has been different for different countries, and for different states within the United States. I will focus first on US drug approval and legal history and then touch on international law.

Clinical Trials

There have been many different kinds of clinical trials over the intervening two centuries since O'Shaughnessy's initial study. What is a modern clinical trial, and why do we put so much effort and money into them? A clinical trial has two main purposes and several minor ones. The first main purpose is to gauge how safe the drug or procedure is. This is usually carried out with an animal model system. Any compound that is slated as a medicinal agent needs to show minimal to no adverse impact on the people it will be administered to. Certain adverse side effects can be tolerated in a drug treatment scenario, but it is important for physicians to know the side effects so that they can evaluate the impact of such effects on an individual basis. For instance, if a side effect of a drug treatment is slightly raised blood pressure, you probably wouldn't choose to administer it to a person with heart problems. On the other hand, this side effect might be tolerable if a patient has no history of heart problems.

Randomization is an important factor in a clinical trial. If a previous treatment exists, the efficacy and safety of a new drug can be tested against the existing one. In this kind of trial, the participants in the trial are separated into controls (those taking the older treatment) and the trials (those taking the new treatment). The assignments are done either double-blindly or "open-label." Double-blind simply means that neither the clinicians doing the testing nor the subjects of the clinical test know the source of the treatment (that is, whether it is the older or newer treatment). After the collection of data relevant to the effect of the two treatments are collected, the identity of the treatment is revealed for analysis. Open-label means that both the clinician and patient know the kind of treatment being administered.

Randomization removes any bias a clinician might have introduced into the trial, and so it is usually the best approach to a clinical trial, although many clinical trials early in the development of a drug are designed as open-label. A second kind of trial involves the use of a placebo, which is a treatment that should have no physiological, neurological, or disease-related effects. The trial treatment is

Table 14.1. CLINICAL TRIAL DATA FOR DRAVET SYNDROME

Variable	Cannabidiol	Placebo	Median Difference	P-value
Convulsive seizures per month				
Baseline	12.4	14.9		
Treatment	5.9	14.1		
Percentage change seizures	−38.9	−13.3	22.8	0.01

then tested for efficacy and safety against the placebo, and again the treatments (placebo and trial) are administered in a double-blind fashion. Quantitative data are best for these kinds of trials, as they are the best suited to statistical testing. The statistical testing usually tells the clinician how different the results are compared to randomness (called significance). If the significance level is set at less than 5 percent (which means that the results would only be obtained 5 percent of the time if the system was completely random), then the treatment is deemed to be better than random 95 percent of the time. This statistic has what is called a *P*-value of less than 5 percent.

The statistical test is a good one, as one wants to know whether there is a real trend toward an effect with the administration of a novel treatment, and using randomness as a benchmark is fully appropriate in this context. As an example, consider the results of the first double-blind placebo clinical trial conducted by neurologist Orrin Devinsky and colleagues, who tested 120 children with Dravet syndrome, a debilitating disorder involving epileptic seizures. Table 14.1 shows the results of the clinical trial. Placebo treatment resulted in nearly the same number of convulsive seizures per month (about 14) as a baseline estimate of number of seizures. But cannabidiol treatment reduced the number of seizures almost 40 percent (from the 12.4 baseline to 5.9 in treated children). The *P*-value estimating the difference between this result and random is significant at less than 1 percent, meaning that the result would only occur 1 percent of the time if the treatment had a random impact. This is strong evidence that the treatment is much better than random. In the same study, Devinsky and colleagues showed that the treatment had little impact on side effects such as sleep disruption, quality of life, and hospitalizations.

The Bottleneck

There are four formal phases to a clinical trial. Each of these phases is examined by the Food and Drug Administration (FDA), which has the authority to pass the drug or procedure to the next phase and ultimately to approve the treatment or drug. Phase I trials involve either animal systems or healthy humans (twenty to one hundred people), to judge how safe the treatment is and to settle on proper dosage for the proposed drug or procedure. Once the FDA review passes the drug or procedure on Phase I, the trials go to Phase II, which emphasizes tests on effectiveness—in other words, whether the drug does what its developers say it does. Designing a Phase II trial is tricky, because the data collected need to convince the FDA that the drug or treatment is effective. Usually between 100 and 500 people are used in these trials, and they can take up to several years to complete. Safety and long-term side effects are other issues that are expanded upon in Phase II trials. Phase III trials involve an order of magnitude more people (about 3,000), enabling the FDA to assess the impact of the drug on different groups of people, different dosages, and using the drug with other medications. If the FDA deems the treatment or compound as passing Phase III, it is okayed for use by the general public. But the full trial process isn't truly over, as a complete assessment of a drug's efficacy and safety can only be addressed with very large sample sizes and longer-term analyses of side effects, which can only be assessed with manufactured and approved compounds. This latter process, called Phase IV, remains an ongoing process even after the compound or procedure has been approved by the FDA.

On average it takes 10 to 15 years for a compound or treatment to pass Phase III, after which the item then is manufactured and used in commonplace medical decisions. A typical trial to bring a new drug to market costs between $1 and 2 billion. Figure 14.1 outlines the pharmaceutical drug development cycle and shows the success rate at different phases of clinical trials. Because drug development is so expensive, some significant and sometimes alarming new trends have been observed. The number of compounds passing at each phase is getting smaller and smaller. Only 40 percent of all compounds tested even make it out of Phase I and are passed on to Phase II, and one-third of these advance to Phase III, with only one-third of these finally being approved by the FDA. If you do the math (one-third of one-third of two-fifths), only about 5 to 10 percent of any compounds entering clinical trials will end up being approved. This creates a huge fi-

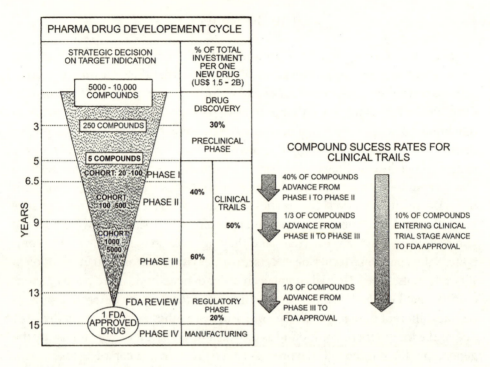

Figure 14.1. Clinical trial sequence, as explained in the text. Data from Harrer et al. (2019).

nancial problem for drug companies. Most of the expense for development and research for a clinical trial occurs later in the process, after close to 60 percent of the funds for a drug approval have been spent (Phase II and Phase III are particularly expensive).

  If a compound doesn't pass Phase III trials, there can be a huge financial loss for the drug developer (some estimate that the loss will be around $1 billion). In fact, the number of new drugs approved by the FDA has halved about every nine years. Those readers familiar with data processing might remember Moore's law (it involves the numbers of transistors on microchips, but can be summarized as the expectation that computing speed and capability will double about every two years). What is happening with drug development is the opposite, resulting in what is called Eroom's law, essentially the opposite of Moore's law: the number of new drugs that are made available halves every nine years or so. This is an alarming problem that begs for a revised clinical trial process to cut the costs of clinical trials. Cost cutting would also presumably result in quicker trial times. Such changes would enhance the development of cannabis-related drugs for approval by the FDA.

The Test Case to End All Test Cases

In fact, we already have a great test case, in which a drug was developed at so-called warp speed: the COVID-19 vaccines. As Anthony Fauci has stated, "The development of several highly efficacious vaccines against a previously unknown viral pathogen, severe acute respiratory syndrome coronavirus 2 (SARS-CoV-2), in less than 1 year from the identification of the virus is unprecedented in the history of vaccinology." Understanding how the vaccines for COVID-19 were expedited might give us some insight into potential expedited cannabis approval.

Most people believe that the development of the COVID-19 vaccines was triggered by the pandemic, and that actual work toward vaccines started with the sequencing of the SARS-CoV-2 genome during the pandemic. It then took less than a year to win approval for the first vaccines. This notion that the vaccines were developed at lightning speed and with insufficient care has contributed to vaccine hesitancy and is used by anti-vaxxers as a reason to distrust the vaccines. As Fauci has pointed out, though, there was a lot of previous science extending back a decade that was involved in the development of the vaccines.

There is a basic misunderstanding among the public that scientists start from ground zero when they develop a vaccine. But the basic science that goes on daily in laboratories across the globe contributes greatly to any and all advancements in medicine and especially vaccinology. Fauci observes that "two activities predate the successful COVID-19 vaccines: the utilization of highly adaptable vaccine platforms such as RNA (among others) and the adaptation of structural biology tools to design agents (immunogens) that powerfully stimulate the immune system." Using these important principles, it was straightforward to design a vaccine. Because of a previous SARS outbreak and a previous MERS outbreak, the core technology had been honed and tested in early phase trials. Indeed, design of vaccines in the future will continue to benefit from these two building blocks. All that was needed was the DNA sequence of a target molecule that our immune systems could focus on. When the genome of SARS-CoV-2 was generated, researchers highlighted the gene sequence of the spike protein of the virus, which attaches it to our cells during infection.

The rapid development of the vaccine was not the only part of drug development that needed to be accelerated. The development of the vaccine coincided with a huge spike in frequency of infection with the virus. This meant that clinical trials for COVID-19 could easily recruit large sample sizes that most clinical trials cannot match. Indeed, COVID-19 sample sizes provided 90 percent efficacy estimates for Phase III tests in less than a year because of the large number of people enrolled in the studies. FDA regulatory steps were easily passed because of the safety of mRNA and viral vector vaccine platforms established well before the pandemic, and the efficacy of the specific COVID-19 vaccines established in Phase III.

Is it possible to speed the adoption of cannabis-derived drugs like the SARS-CoV-2 vaccines demonstrated? The most time-consuming and costly step in drug development is Phase III testing. Can Phase III testing be compressed for cannabis? It remains to be seen, but there are other ways to move cannabis-related pharmaceuticals through the approval process.

Step by Step

Because the approval process for cannabis as a medicinal product and the commercialization of such a product are both complicated, let's review how cannabis or a cannabis-derived product would move through an FDA approval process. First, we need to make a distinction between cannabis products from the plant itself (cannabis-derived) and cannabis synthetics or derivatives that in one way or another mimic natural cannabinoids. The approval guidelines for compounds in these categories differ from each other. Because the synthetics involve inherently different kinds of chemicals than naturally occurring ones, the regulations for these two kinds of cannabinoids are different. Sometimes a synthetic can cross lines, such as synthetically derived dronabinol that also occurs naturally in the cannabis plant. Compare this to the synthetic cannabinoid nabilone, which is not naturally occurring.

The difference between synthetic and naturally occurring cannabinoids should be clear. How they are treated during the approval process differs as a result of the THCA concentration in the drug being tested. Currently, as a result of the US Agriculture Improvement Act of 2018, any cannabis product containing THCA at a concentration higher than 0.3 percent is considered a Schedule 1 narcotic, while cannabis products under 0.3 percent THCA are not Schedule 1, so there are two pathways a cannabis-derived pharmaceutical can follow.

The approval of a novel pharmaceutical starts with a sponsor (usually a pharmaceutical company) that has developed the drug. For the products with less than 0.3 percent THC, the steps are:

1. The sponsor obtains a pre-investigational new drug (IND) number through the FDA Center for Drug Evaluation and Research (CDER). At this point the sponsor can request a meeting with the FDA to explain the drug and ask for guidance. This part of step 1 is optional.
2. The sponsor explains to the FDA all the relevant chemistry, manufacturing, and controls (CMC) if the product is a synthetic, or botanical raw material (BRM) information if the product is a botanical. This step is reviewed by the CDER.
3. The third step involves a mechanism the FDA has developed called a Drug Master File (DMF). These are files describing the nitty-gritty of drug manufacture that the sponsor can use in the application. There are nearly 38,000 of these master file subjects that the FDA has considered. The DMFs make the application less redundant if previous technology or chemistry is part of the novel drug. Safe use in humans is the major focus of this step, and if a DMF is not used then the sponsor has to provide information on safe use in humans.
4. The IND and clinical protocol are officially submitted to the FDA.
5. The FDA reviews the submitted IND. There is a thirty-day waiting period before the sponsor can start clinical trials (unless the FDA allows it sooner or denies the IND).

For a cannabinoid pharmaceutical product with THC content greater than 0.3 percent, the steps are basically the same, except that additional ones are included between step 1 and step 2, and there is an additional step at the end. Since the developed pharmaceutical by definition involves a Schedule 1 drug (it has a concentration of THCA greater than 0.3 percent), the additional steps require notifying the Drug Enforcement Administration (DEA) or the National Institute on Drug Abuse (NIDA) that the application involves a Schedule 1 narcotic. Before the last step, the FDA informs the sponsor that the IND is authorized as "safe to proceed"; then the sponsor can apply to the DEA to obtain a protocol registration that if granted allows the sponsor to obtain the controlled substance from the NIDA or DEA. The sponsor can then begin the clinical trial.

While this procedure might seem a bit cumbersome, the FDA has also instituted some work-arounds to speed up the process. The FDA for a long time has recognized the need for rapid processing through these steps, but it also aims to maintain a balance between safety, getting the regulatory steps right, and speed. Several options exist to expedite the process: Fast Track, Breakthrough Therapy, Accelerated Approval, and Priority Review. The FDA has recognized that plant-derived drugs are also in need of approval processes and recently established a Botanical Review Team that purports to assist sponsors in rapid review of botanically derived products such as cannabis. Douglas Throckmorton, the former deputy director of the Center for Drug Evaluation and Research at the FDA, was clear in stating that

> the misperception persists that all products made from or containing hemp, including those made with CBD, are now legal to sell in interstate commerce. The result has been that storefronts and online retailers have flooded the market with these products, many with unsubstantiated therapeutic claims. FDA has seen CBD appear in a wide variety of products including those purporting to be foods, dietary supplements, veterinary products, and cosmetics. As this new market emerges, we have seen substantial interest from industry, consumers, and Congress. However, FDA's role remains the same: to protect and promote the public health. At present, any food containing CBD or purported CBD dietary supplement product in interstate commerce is in violation of the FD&C (Food, Drug and Cosmetic) Act.

Specifically, the products that Throckmorton addresses are illegal if not having FDA approval. When producers of these so-called illegal cannabis products advertise them as medicinal, the FDA has taken the tactic of sending the offending company or individuals a "warning letter" instead of prosecuting. There has been an increase in the number of warning letters since 2015, when no letters were issued. In 2019 nearly twenty letters were issued, and in 2022 nearly thirty. This trend is somewhat alarming, indicating that more cannabis companies are making claims about their products that they should not, because they do not have FDA approval.

The State of the Art

To date, the FDA has not approved a marketing application for cannabis plants for the treatment of any disease or condition. The agency has, however, approved cannabis-derived drug products. Epidiolex has been shown in clinical trials to be effective in the treatment of two specific disorders called Lennox-Gastaut syndrome and Dravet syndrome (discussed earlier in this chapter). In 2014 a United Kingdom–based drug company started Fast Track designation for this drug, leading to three randomized clinical trials with children suffering from the syndromes. Two trials were accomplished, along with the one described earlier by Devinsky and colleagues, on a total of 516 children. All three clinical trials were double-blind placebo-based. The cannabidiol treatment administered (given the commercial name Epidiolex) was highly effective at curbing the number of seizures and was deemed safe. Some mild and infrequent side effects were uncovered, but these, when weighed against the benefit of reducing seizures, were minimal. Given these data, the clinical trial passed FDA approval.

Three synthetic cannabis-related drug products—Marinol, Syndros (both derived from dronabinol), and Cesamet (derived from nabilone)—have also been approved for use but not for marketing. These approved products are only available with a prescription from a licensed health care provider. The FDA has not approved any other cannabis, cannabis-derived, or cannabidiol (CBD) products currently available on the market.

There are dozens of other synthetics. Table 14.2 lists the seven major categories of synthetic cannabinoids. These categories are based on the basic core chemical structure of the synthetic compound. Common names for synthetics in general are herbal incense, potpourri, fake weed, legal weed, K2, Spice, Gold Spice, Silver Spice, Diamond, Bliss, Black Mamba, Bombay Blue, Blaze, Genie, Zohai, JWH-250, Kronic, Yucatan Fire, Skunk Moon Rock, AK-47, Mr. Happy, Scooby Snacks, Kush, and so on. Not regulated and generally sold over the counter, synthetics have gotten a deserved bad name.

They are for the most part dangerous because they are not regulated, and most have never come even close to receiving clinical trials. Synthetics can be up to 100 times more concentrated with psychoactive compounds, and because sometimes a compound is made in different labs, the batches are often inconsistent in concentration, purity, and quality. On the other hand, they have been packaged slickly, and many have "not for human consumption" stamped on the packaging. The early 2000s saw the heyday of synthetics such as K2 and Spice, which were marketed and sold openly. But in 2011, as a response to growing reports of adverse effects of these synthetics, the DEA decided to schedule the chemicals in synthetics as Schedule 1 narcotics. This emergency move by the FDA led to President Obama's signing the Synthetic Drug Abuse Prevention Act in 2012. This

Table 14.2. SEVEN MAJOR CHEMICAL CATEGORIES OF
SYNTHETIC CANNABINOIDS

1. Naphthoylindoles (e.g., JWH-018, JWH-073, and JWH-398)

2. Naphthylmethylindoles

3. Naphthoylpyrroles

4. Naphthylmethylindenes

5. Phenylacetylindoles (i.e., benzoylindoles, e.g., JWH-250)

6. Cyclohexylphenols (e.g., CP 47,497 and homologues of CP 47,497)

7. Classical cannabinoids (e.g., HU-210)

Note: Names such as Napthoylindoles are a mouthful, and so abbreviations
are often used to refer to these chemicals. Those with JWH (such as
JWH-018) refer to synthetics made by one person's lab: John W. Huffman, a
chemist who synthesized over 400 different cannabinoids during his career.

act permanently moved many of the cannabis-related synthetics' chemical compounds onto Schedule 1 lists of the Controlled Substances Act (CSA; see chapter
15 for a discussion of the CSA schedule system).

The number of FDA approvals to date is minimal, given the large number
of cannabis strains, natural cannabinoids, and synthetics and the many potential
medical uses that have been documented anecdotally. The snagging point is the
clinical trials determining whether a substance is safe and efficacious. In 2000 the
National Institutes of Health initiated an online listing of what clinical trials have
been done, which have been terminated before conclusion, and which have been
complete. This website is called ClinicalTrials.gov (see For Further Reading), and
nearly a half million clinical trials in both the US and 221 other countries are listed
in the database as of the writing of this book. A search of the database using the
word "cannabinoid" yielded 564 clinical studies. Not all of these are studies about
the safety or efficacy of cannabis. Some clinical trials are focused on drugs that
will help with cannabis dependence, and there are many disorders where only a
few clinical trials are reported in ClinicalTrials.gov. Figure 14.2 shows several of
the most studied disorders involving cannabis therapies. Obesity tops out with
thirty-one individual clinical trials focused on that disorder, followed by general
pain (twenty-seven), schizophrenia (twenty-two), cancer-related pain (18), and
PTSD (16).

With this large number of clinical trials either planned or completed, we can

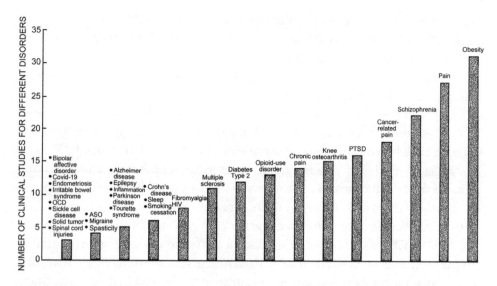

Figure 14.2. Cannabis clinical trials listed by www.clinicaltrials.gov with three or more studies as of 2022. The bars represent the number of individual trials undertaken on the indicated disorder. For instance, eight studies have three clinical trials each, and obesity has been studied in thirty-two clinical trial studies. There are over 100 disorders that either address cannabis dependence or cannabis dosage or have only two or fewer trials. These have all been left out of the figure. Data were taken from www.clinicaltrials.gov. Different results might arise with different search terms and alternative ways of sorting disorders, but the general trends in the figure illustrate the different focal points of clinical trials with cannabis-derived compounds.

probably expect some movement by the FDA to approve cannabis for some of these disorders. Of course, approval through the clinical trials also depends on the results and quality of the trials.

Entourage Effects and Medicinal Cannabis

There are further challenges involved in the development of medicinal cannabis beyond the clinical trials and legal issues. The approach to drug development can focus on single synthesized molecules, on molecules extracted from the plant itself, or by using the whole plant. There are advantages to each of these approaches. But as anyone knows who has used marijuana and explored even a few of the many hundreds of different strains, imbibing the natural product without knowing about the strain can be an adventure. Researchers Dvora Namdar, Omer Anis, Patrick Poulin, and Hinanit Koltai point out that "patients should have confidence in the drug's uniformity, strength, and consistent delivery that support appropriate dosing needed to treat patients with complex and serious conditions."

Cannabis is a remarkably complex plant with respect to the compounds it makes, and some researchers believe that these compounds interact with each other to produce its medicinal and recreational results. With this "entourage effect" some noncannabinoid compounds may potentiate the impacts of THC and CBD on the human body. The father of cannabis compound research, Raphael Mechoulam, pointed to this entourage effect as one of the most important new areas for medicinal cannabis researchers to investigate: "Although this is not observed in all effects of THC or all effects of cannabinoids, many of them have this 'entourage effect' and so things become very complicated; if somebody wants to put THC or CBD as an extract, they also need to know what other compounds are present." A dose of purified CBD therefore might not be as efficient at stimulating an endocannabinoid system response in the human body as a treatment that includes terpenes or other secondary compounds of the plant. Namdar and colleagues point to several additional entourage effect considerations. First and most obvious is that other compounds could also retard or damp down the impact of a cannabis-derived drug, a syndrome called the "parasitage effect." Such compounds should be inactivated or removed from any cannabis-derived medicinal to better control the drug's impact. Second is the possibility that the endocannabinoids our bodies make might interact or interfere with the cannabinoids that come from the plant.

This seems to be a critical aspect of drug development that should not be overlooked, especially if treatment with a cannabinoid-derived compound might seriously disrupt natural endocannabinoid function. Deciphering these kinds of interactions would lead to a better understanding of the side effects caused by treatments with cannabis-derived pharmaceuticals. Finally, we need to understand the degradation and turnover of pharmaceutical cannabinoid products better, in order to manage dosage and side effects more effectively. Understanding the entourage effect and how to mix and match compounds will be a critical part of drug development of cannabinoid-based compounds. Making the development of drugs from cannabinoid-related products safer and more controlled is a requirement for their continued use in medicine. Some of the effects of cannabis are considered to be dangerous and addictive, and therefore make the use of cannabis-derived pharmaceuticals complicated. The potential dangers of cannabis are discussed further in chapter 16.

15

Legalize It?

In 1973 during my second year at college, I met my best college friend, who I believed was destined to succeed at everything he ever tried. He eventually wrote briefs submitted to the US Supreme Court on the liberal side of several legal subjects. When I told him about this book, he was eager to brief me on the legal history of commerce's regulation in the United States. His take strongly influenced how I try to tell the story here.

He pointed out to me that it wasn't until the twentieth century that the federal government became interested in regulating any kind of products, either at the civil or the criminal level. State governments traditionally did this. Two big changes, however, stimulated the federal government's intrusion: railroads and corporations. Railroads disdained state governments' regulation of their industry. Because they often spanned several states, railroads wanted a more uniform set of rules. They also wanted a system they could control more easily. My friend noted that "federal regulation would preempt state regulation, and knowing that they could control what the federal regulation was and that the regulators would work with a very light hand, they called for it."

The role of large corporations such as oil companies was somewhat different. As progressivism began to influence how Americans viewed their society, progressives called for tighter regulation of corporations to rein them in. In the early part of the twentieth century, the Supreme Court contested the federal legality of some of the regulatory measures passed by Congress. The Court was hostile to many attempts at federal regulation (including minimum-wage laws, child labor laws, and agricultural relief laws) until President Franklin Delano Roosevelt threatened to pack the Court with additional justices. The Court then decided that endeavors that affected interstate commerce were legitimately the subject of federal regulation. The civil rights era justices continued to think in this way, and only recently has there been a move away from federal regulation of many things. My lawyer friend calls this "the background against which the hodge-podge of which current state and federal regulation occurs." Indeed, it has had a huge effect on the legal context for cannabis commerce.

Americans have seen a loosening of state regulations so far in the early 2020s, while there have been few stirrings at the federal level. This situation could change as the federal landscape shifts. Congress could very well act to modify the regulation of cannabis, but so far it hasn't. This all occurs in a strange political environment, in which as my friend put it, "left-leaning liberals and right-leaning libertarians have joined forces to push through state legislation to make cannabis more accessible."

Another Alphabet Soup

Cannabis was a fairly well-known quantity in the early history of the United States, and even more so as the twentieth century came and went. However, temperance movements and societal recognition that access to opiate drugs needed to be controlled gradually became more popular. People in government and those interested in temperance increasingly advocated for regulation. The first drug regulation law in the United States was the 1906 Pure Food and Drug Act, which focused mainly on opiates. In 1909 the US Congress passed the Act to Prohibit the Importation and Use of Opium, which did what its title suggests. The first US law against marijuana was a 1914 regional law in El Paso, Texas, which banned the sale and possession of cannabis in that border town. In addition to being the first US ordinance against marijuana, it established a strategy for creating laws against cannabis use based on a theory of criminality: marijuana was used by shady people and hence enhanced criminality. Some of the wording in the ordinance is quite racist and classist (marijuana is used by "Negroes, prostitutes, pimps, and a criminal class of whites"). This ugly theory unfortunately endured past the early part of the twentieth century and to a large degree continues to live on in the way the perception of criminality of using marijuana (and other drugs) is applied to citizens of the United States today.

A federal act was passed by Congress in the same year as the El Paso ordinance, called the Opium and Coca Leaves Trade Restrictions Act. It was sponsored by a New York Democrat named Francis B. Harrison and became better known as the Harrison Narcotics Tax Act. It required that a tax be paid by any importer or user of opium or cocaine and was initially passed as a revenue-focused act. Here is where narcotics law gets tricky. Cannabis industry lawyer David B. Patton points out that "as American government was understood in 1914, only States could regulate the practice of medicine, and the federal government could not do so unless the activity involved interstate commerce. But the federal government expressly has the constitutional power to tax." So the burden of regulating controlled substances fell on the shoulders of the federal government. While the Harrison Act started out as a revenue-controlling law, it eventually turned into a temperance law, and states began to pass "little Harrison Acts" that regulated opium and cocaine in individual states. In 1922 the US Congress passed the Narcotic Drugs Import and Export Act. This act criminalized narcotic possession and established the Federal Narcotics Control Board, which became the Federal Bureau of Narcotics (FBN) in 1930. Several decades later (from 1966 to 1968), the Bureau of Drug Abuse Control (BDAC) emerged as a part of the Food and Drug Administration. In 1968 the FBN and BDAC were combined into the Bureau of Narcotics and Dangerous Drugs (BNDD). And in 1973 the BNDD and several other small government agencies involved in narcotics regulation were combined into the Drug Enforcement Administration (DEA) within the Justice Department. It is ironic that the regulatory agency in charge today started out as a Treasury Department agency, morphed into a Food and Drug agency, and then was subsumed by the Justice Department.

While this alphabet soup of agencies was settling in the soup bowl of the Department of Justice, new legislation was passed and new leaders were appointed who further regulated narcotics in the US. Harry Anslinger was appointed the head of the FBN in 1930 and served well into the 1960s. He was no fan of narcotics and of marijuana specifically; at one point in 1937 he commented that "if the hideous monster Frankenstein came face to face with the monster Marihuana, he would drop dead of fright." Anslinger was head of the FBN when the Uniform State Narcotic Drug Act was passed in 1932, which was the first time cannabis was classified as a legally defined narcotic. Because only ten states adopted the law at first, he started an aggressive propaganda program highlighting the risks of marijuana use. He enlisted various groups such as the Woman's Christian Temperance Union, the General Federation of Women's Clubs, and the National Catholic Welfare Council to battle the "dangers of marijuana." His supporters became known as "Anslinger's Army." When one thinks of any anti-marijuana propaganda, *Reefer Madness* should come to mind, and indeed this 1936 film (also titled *Tell Your Children* and *Doped Youth* in later incarnations) was promoted by Anslinger's FBN as the gospel truth about the perils of marijuana use. (Nothing but bad happens to those who use cannabis in the film.)

Oddly enough, a Supreme Court ruling on gun taxes was the next major influence on the legal status of marijuana. In 1934 the National Firearms Act was passed, requiring gun sellers to pay a $200 tax on every gun they sold. It was intended as a prohibitive tax (the legislators hoped that the sale of guns would be slowed by the tax). A Mr. Sonzinsky was caught avoiding the tax and was prosecuted. His lawyers took the case all the way to the Supreme Court on the basis of the tax being unconstitutional. The Court supported the law as a prohibitive one and not a revenue-raising tax statute; the tax was levied to curtail the behavior of citizens and not to make money for the government. With that decision on the books, the US Congress and Anslinger moved to make marijuana even more difficult to sell and use by passing the 1937 Marihuana Tax Act. The pharmaceutical industry reacted negatively to this tax because it would in effect eliminate any research on marijuana as a medicinal. Indeed, medicinal marijuana became nonexistent from the 1940s to the latter part of the century. This single tax act and Anslinger's aggressive politicking would retard the growth of knowledge about medical marijuana for over a half century. Worse, the many positive aspects of marijuana were buried under the classification of cannabis as a dangerous narcotic.

The one dissenting voice in the battle against Anslinger's Army was a 1944 commission in New York City that produced a report entitled "The Marihuana Problem in the City of New York" (known as the La Guardia Report). This assessment commissioned by Mayor Fiorello La Guardia strongly debunked the criminality theory. The commission concluded that

- The practice of smoking marihuana does not lead to addiction in the medical sense of the word.
- The use of marihuana does not lead to morphine or heroin or cocaine addiction and no effort is made to create a market for these narcotics by stimulating the practice of marihuana smoking.

- Marihuana is not the determining factor in the commission of major crimes.
- Juvenile delinquency is not associated with the practice of smoking marihuana.
- The publicity concerning the catastrophic effects of marihuana smoking in New York City is unfounded.

While the La Guardia Report should have been taken as a message to back off, further anti-cannabis legislation was passed in the 1950s. The Boggs Act of 1951 set out uniform punishment regimes for narcotics laws violation. Because cannabis was considered a "gateway" drug, it was classified with dangerous opiates and other hard drugs. This resulted in cannabis users and sellers having draconian punishments levied against them if they were caught from the 1950s through to the 1980s. Through the rest of the 1950s, many state governments created several "little Boggs Acts."

In 1969 Timothy Leary of LSD fame challenged the federal marijuana tax act and won his case (*Leary v. United States*) in the Supreme Court. This decision rendered the tax act unconstitutional. This case and changing mores regarding drug use in the 1960s and 1970s triggered a doubling-down response from the federal government. In 1970 the Comprehensive Drug Abuse Prevention and Control Act and the Controlled Substances Act (CSA) were passed by the US Congress. The CSA also introduced the concept of "schedules" as descriptors of narcotic danger or capacity for abuse. According to the law, there are five tiers or schedules regarding narcotics. Schedule 1 narcotics are the most threatening, with high potential for abuse and no medical uses or accepted safety protocols for the drugs, even under medical supervision. Heroin and LSD are two of the more visible narcotics in this category. Schedule 2 are drugs that have medical utility but are dangerous because they have a high probability for abuse; these include codeine, morphine, and methadone. Schedule 3 are drugs with lower potential for abuse than Schedule 2. There is a low probability of being physically addicted to them, but a high possibility of psychological dependence. An example of this schedule drug is Tylenol with codeine. Schedule 4 are drugs that have a low propensity for abuse compared to Schedule 3 drugs. They are still regulated because even a low probability of addiction is enough to require caution. Drugs such as Xanax, Valium, and clonazepam are in this schedule. The final category—Schedule 5—included drugs that have lower propensity for abuse compared with Schedule 4. A prescription drug such as Robitussin AC is an example of this category.

These definitions are rather fuzzy, and it is difficult to discern when there might be a switch from one schedule to another. Indeed, some individuals have argued that drugs that are listed in one particular "schedule" might possibly better belong in another. When we categorize phenomena in science, there is much more precision and objectivity in what a scientist uses to assign one thing to this category and another thing to that category. Arguing that one schedule has "potential for abuse less than substances in a previous category" relies on the word "less," and that word is not precise for the potential for abuse. Marijuana has been caught up in the subjectivity of the schedules scenario for some time.

The CSA gave the US attorney general the power to assign various drugs to particular schedules, and marijuana was initially categorized in the scariest tier as a Schedule 1 narcotic. Many at the time (1970) objected to the listing of marijuana as a Schedule 1 narcotic, and several introduced legislation to move it off. Then "Tricky Dick" got involved.

The Battle of the Attorneys General

In 1972 Richard Nixon, then president of the United States, made it a personal crusade to bring a hammer down on marijuana. In a jewel of twisted logic, he cemented the listing of marijuana as a Schedule 1 drug with this quote: "In recent days, there have been proposals to legalize the possession and use of marihuana. I oppose the legalization of the sale, possession or use of marihuana. The line against the use of dangerous drugs is now drawn on this side of marihuana. If we move the line to the other side and accept the use of this drug, how can we draw the line against other illegal drugs? . . . There must continue to be criminal sanctions against the possession, sale or use of marihuana."

Without considering the medicinal utility of marijuana, Nixon drew a subjective line in the sand over which he would not let others cross. Marijuana therefore continued as a Schedule 1 drug, but its use by the public continued to broaden and attitudes toward it began to shift. Something had to give, but it took a few decades. Post-Nixon legal maneuvers have occurred through the administrations of nine succeeding US presidents (Ford, Carter, Reagan, Bush I, Clinton, Bush II, Obama, Trump, and Biden). For instance, in 1980 Ronald Reagan's attorney general Benjamin Civiletti set forth specific guidelines for the prosecution of marijuana offenses. These guidelines were basically followed by the US Department of Justice (DOJ) for the next thirty years.

Small steps toward legalization occurred during some presidential administrations, followed by setbacks in others. The first state to move toward legalization was California, which in 1996 legalized the use of medical applications of
marijuana. Several states followed suit, but when California took the first big step
the federal government stepped in. The "drug czar" at the time, Gen. Barry R.
McCaffrey, issued a joint statement with the DOJ, DEA, and HHS (Health and
Human Services), announcing that any doctor who prescribed a Schedule 1 drug
was not acting in the public interest and that "such action would lead to revoca-
tion of the physician's registration to prescribe controlled substances." Even
speaking to a patient about the potential benefits of a Schedule 1 drug such as
marijuana would result in the suspension of the physician's controlled substances
registration certificate, issued by the DEA. Without this certificate a doctor could
not prescribe a drug; without the capacity to write prescriptions, a doctor would
be rendered almost useless.

In other words, you prescribe a Schedule I drug, we will ruin your medical
career. The director of the Office of National Drug Control Policy and the US
departments mentioned above were sued by several patient and doctor groups
in the early 2000s. The case, known as *Conant v. Walters,* was decided by review
in an appellate court. The court ruled that the stripping of DEA's controlled substances certificate because of discussion with patients about marijuana as a medicinal was a violation of the free-speech clause of the US Constitution. The case
was brought to the Supreme Court, but the Court refused to review it, so the appellate court decision stood pat in 2002. But prescribing, providing the prescription, and possessing the prescription still remained illegal. It was a simple step for
the medical community to use the word "recommend" instead of "prescribe" and
the forty-four states (plus the District of Columbia) that subsequently legalized
medicinal cannabis have lived with this terminology.

There was still a long way to get where we are now, with twenty-four states
(and the District of Columbia, plus three American territories) approving fully
legal use of medicinal marijuana. A lot of legal wrangling still needed to happen
to get these states to the legalization stage. In 2009 Eric Holder, Barack Obama's
first attorney general, asked his deputy attorney general David W. Ogden to address the legality of marijuana; by then sixteen states had legalized medicinal
marijuana, and the time seemed ripe for broadening the legality of cannabis. The
so-called Ogden memo clearly stated that it would be a waste of time and money
for federal law enforcement to prosecute patients, doctors, and other caregivers
who were lawfully allowed to prescribe, sell, buy, and use marijuana if the use fell

Table 15.1. THE DON'T DO'S OF THE SECOND COLE MEMORANDUM

1. Preventing the distribution of marijuana to minors

2. Preventing revenue from the sale of marijuana from going to criminal enterprises, gangs, and cartels

3. Preventing the diversion of marijuana from states where it is legal under state law in some form to other states

4. Preventing state-authorized marijuana activity from being used as a cover or pretext for the trafficking of other illegal drugs or other illegal activity

5. Preventing violence and the use of firearms in the cultivation and distribution of marijuana

6. Preventing drugged driving and the exacerbation of other adverse public health consequences associated with marijuana use

7. Preventing growing marijuana on public lands and the attendant public safety and environmental dangers posed by marijuana production on public lands

8. Preventing marijuana possession or use on federal property

Source: Patton (2020).

under the umbrella of sanctioned medical marijuana use. The Ogden memo did not sanction the use of marijuana if it was involved in the unlawful use of firearms, if it was involved in or caused violent criminal activity, if it was sold to minors, if there were any unlawful money dealings involved in its sale or use, if a person had it in their possession more than was needed for medical use, if marijuana was involved in a crime involving other Schedule 1 narcotics, and if it was part of an organized criminal enterprise.

Another deputy attorney general issued two memos after the Ogden memo. James M. Cole issued the first memo in 2011. It reconfirmed the nuances of the Ogden memo but focused on large clandestine marijuana growers, recommending that such growers should be a priority for prosecution by the DOJ. The second Cole memo came in 2013 during Obama's second term, when twenty-one states had legalized medical cannabis. It was more of a "don't do this" list. The DOJ hoped to prevent certain over-the-top aspects of marijuana commerce (table 15.1) and promised to prosecute those who broke their "don't do" list.

While the second Cole memo allowed the DOJ to prosecute marijuana use in specific cases, it was basically a passive ticket to prescribe, sell, and use marijuana for medicinal purposes. The management of cannabis commerce was basically left to the states that had legalized it. These memos suggested that prosecution by the DOJ would not be undertaken if people stayed within their "don't do" list; but legally the DOJ could still prosecute anyway.

With the 2016 election of Donald J. Trump as president came the appointment of Jefferson Beauregard Sessions III, who hated marijuana with the same fervor as Anslinger had seventy years earlier. In 2017 Sessions issued a memorandum rescinding the Ogden memo and the two Cole memos. In the opposite spirit, though, the US Congress passed the Agriculture Improvement Act of 2018, aiming to control the growing of marijuana by allowing "hemp" to be grown on farms. The tricky factor here was what "hemp" is; Congress defined it in the following way: "The term 'hemp' means the plant *Cannabis sativa L.* and any part of that plant, including the seeds thereof and all derivatives, extracts, cannabinoids, isomers, acids, salts, and salts of isomers, whether growing or not, with a delta-9 tetrahydrocannabinol [THC] concentration of not more than 0.3 percent on a dry weight basis."

Of course, this definition was based on the impact of the psychoactive properties of THC (which lawmakers wanted to limit) versus the nonpsychoactive properties of CBD (which lawmakers didn't want to disrupt, because it had become big business). Hemp growers were thus considered different from marijuana growers, with hemp growers doing legal farming and marijuana growers not.

Sessions was determined to return the status of marijuana use to Civiletti's 1980 guidelines, which included harsh punishment for all prescriptions, sales, and uses of marijuana. It looked like a new dark age was approaching for marijuana use in the US. David B. Patton summarizes four reasons why the dark age did not dawn. First, the Sessions memo backfired on itself, as it left it up to individual US attorney offices to decide whether or not to prosecute a given case involving marijuana. This meant that states such as Colorado, with a lucrative cannabis commerce and taxation in full swing, could decide not to prosecute most cases. Most US attorney offices simply fell back on the guidelines they had been following since the Ogden memo was issued. In other words, the Sessions memo had no teeth.

Second, in 2014 Congress had cut funding to the DOJ and the Federal Bureau of Investigation (FBI) for investigations of cases involving medical marijuana. When the Sessions memo came along, Congress indicated that this defunding trend would continue. In 2019 Congress specifically singled out all states that had pro–medicinal marijuana laws as no-prosecution zones for use of funds dispersed to the DOJ and FBI. Third, states continued to pass legalization mea-

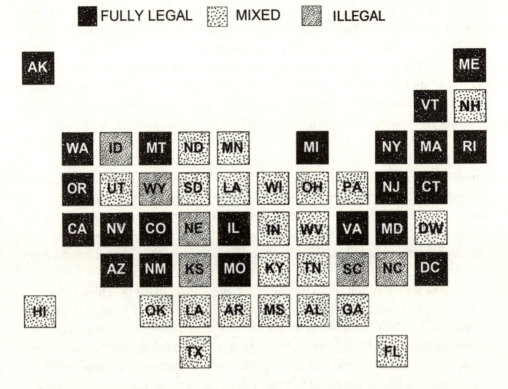

Figure 15.1. Diagram of the United States showing states where marijuana is fully legal (black), mixed legal (light stippling), and illegal (dark stippling). Adapted from https://cfah.org/marijuana-legality-by-state/. This figure was constructed in mid-2025. Marijuana laws change rapidly from state to state. It is always a good idea to check a specific state's marijuana laws before carrying it across state borders.

sures, making it difficult for the DOJ to keep up. This legalization juggernaut is in a very advanced stage now. Over most of the nearly twenty years since California took the legalization plunge in 1996, any map depicting legalization highlighted the states that had legalized it one way or another. Today it is easier and more illustrative to highlight the states that have *not* moved on legalization (fig. 15.1). Six states to date (Idaho, Wyoming, Kansas, Nebraska, North Carolina, and South Carolina) still maintain that marijuana is illegal. Twenty-two states (and the District of Columbia) have made both medicinal and recreational marijuana legal, with the remaining twenty-three states having hybrid legalization laws.

The fourth reason the Sessions memo failed is because of the increased popularity of the marijuana industry. Whenever a strong business interest is involved, laws or application of laws will tend to be favorable to that industry. Cannabis has attained a strong business momentum, evolving from a motley group of independent and secretive growers to a multi-billion-dollar endeavor. As Patton comments, "In the current environment, a large-scale law enforcement crackdown on the cannabis industry would be intolerably economically disruptive and impolitic."

Barr and Biden

Donald J. Trump had more than one attorney general during his first term. The 2019 appointment of William Barr as attorney general put a shock of fear into the marijuana industry because throughout his career he had been strongly supportive of anti-cannabis legalization. He argued that "we should have a federal law that prohibits marijuana everywhere, which I would support myself. Because I think it's a mistake to back off on marijuana." But in a legal turnaround he also stated that to prosecute marijuana offenses as in pre-Ogden and pre-Cole days would be "untenable." He was unwilling to let his personal abhorrence of marijuana influence his legal policy. He also pointed out the conundrum of cannabis use being legal at many state levels but still illegal at the federal level.

Because the attorney general is a presidential appointment, another four years of Republican dominance in the executive branch of the US government might have swung the pendulum away from legalization at both the state and federal levels. But the story changed drastically with the election of Joseph R. Biden as president in 2020. He appointed Merrick Garland as his attorney general, and Garland quickly announced that marijuana prosecution would not be a priority of the DOJ. Biden went a step further with a statement on marijuana from the President's Office on October 6, 2022. It is worth reproducing that statement in its entirety:

> As I often said during my campaign for President, no one should be in jail just for using or possessing marijuana. Sending people to prison for possessing marijuana has upended too many lives and incarcerated people for conduct that many states no longer prohibit. Criminal records for marijuana possession have also imposed needless barriers to employment, housing, and educational opportunities. And while white and Black and brown people use marijuana at similar rates, Black and brown people have been arrested, prosecuted, and convicted at disproportionate rates.

Today, I am announcing three steps that I am taking to end this failed approach.

First, I am announcing a pardon of all prior Federal offenses of simple possession of marijuana. I have directed the Attorney General to develop an administrative process for the issuance of certificates of pardon to eligible individuals. There are thousands of people who have prior Federal convictions for marijuana possession, who may be denied employment, housing, or educational opportunities as a result. My action will help relieve the collateral consequences arising from these convictions.

Second, I am urging all Governors to do the same with regard to state offenses. Just as no one should be in a Federal prison solely due to the possession of marijuana, no one should be in a local jail or state prison for that reason, either.

Third, I am asking the Secretary of Health and Human Services and the Attorney General to initiate the administrative process to review expeditiously how marijuana is scheduled under federal law. Federal law currently classifies marijuana in Schedule I of the Controlled Substances Act, the classification meant for the most dangerous substances. This is the same schedule as for heroin and LSD, and even higher than the classification of fentanyl and methamphetamine—the drugs that are driving our overdose epidemic.

Finally, even as federal and state regulation of marijuana changes, important limitations on trafficking, marketing, and under-age sales should stay in place.

Too many lives have been upended because of our failed approach to marijuana. It's time that we right these wrongs.

                                                          Joseph R. Biden

Biden recognized one crucial aspect of the whole controversy about legalization of marijuana. Over the 100 years or so of criminalization of cannabis, many more lives had been ruined by prosecution and incarceration than those damaged by actual marijuana use. In fact, factoring in the medicinal properties of cannabis truly reveals the inanity of criminalized marijuana. Equally important, Biden recommended the reclassification of marijuana away from being a Schedule 1 drug. As of this writing, nothing legal has happened, but it's no longer a question of if but of when the legal issues surrounding marijuana will be righted.

As the United States Goes, So Does the World?

To review all the different legal issues with every country would be a grueling and tedious process. Figure 15.2 shows the global distribution of cannabis-legal and cannabis-illegal nations. Note that there are several categories of legality and illegality, criminalized and decriminalized, and enforced and unenforced. The laws in other countries are complex and show a great deal of variation. When traveling to a foreign country, it is best to examine the explicit marijuana laws of that country before bringing cannabis with you or purchasing it in the visited country. The recent case of WNBA basketball star Brittney Griner and her incarceration in Russia is a good example of failed detail and care in traveling with marijuana, leading to a traumatic and unnecessary result. Although Griner didn't deserve the medieval gulag sentence she received, a better knowledge of the laws would have kept her out of trouble.

International regulation of narcotics extends back to 1912, when drugs were first addressed by an international commission. The specific international regulation of cannabis started when the 1925 Shanghai International Opium Convention convened. At the insistence of Egypt, cannabis was recognized by this convention as a dangerous drug. While Anslinger's Army was forming in the US, other countries looked to international regulation as an important part of their own regulatory strategies. Historian and policy analyst John Collins points out that contrary to popular belief, the US was not substantially involved in settling international cannabis legal issues in the early days of the League of Nations and later at the United Nations. In fact, the United States (which never joined the League of Nations) walked out of the 1925 Shanghai convention and refused to sign a 1936 antitrafficking agreement sponsored by the League. When the UN developed its Single Convention on Narcotic Drugs in 1961, the US was not supportive. This document attempted to make uniform the illegality of narcotics and formed the International Narcotics Control Board (INCB), which led to an international drug control system (IDCS). The US was somewhat blasé with respect to

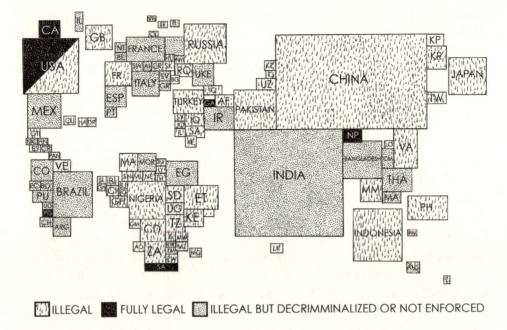

ILLEGAL   FULLY LEGAL   ILLEGAL BUT DECRIMMINALIZED OR NOT ENFORCED

Figure 15.2. Global diagram of country-by-country status of cannabis legality. Data from https://vivid maps.com/cannabis-legality/#google_vignette. This figure was constructed in 2024. Marijuana laws change rapidly. It is always a good idea to check a specific country's marijuana laws before carrying it into another country.

the system and was often absent from decision-making at the global level. Other countries such as Italy, South Africa, Turkey, and Egypt were more involved in the workings of the global legal system.

Since 1961 the INCB pushed forward by more precisely defining cannabis and cannabis products. The term "Indian Hemp" was dropped, and "cannabis" was inserted to be more precise when international issues arose. The INCB urged a distinction between cannabis, cannabis plants, and cannabis resin and recommended that these aspects of marijuana be regulated much in the same way opium was regulated. This led to research on marijuana being regulated the same way opium was. There was also a distinction made between cannabis and cannabis resin versus cannabis for industrial purposes, placing industrial uses in a different regulatory category than medicinal and recreational cannabis and cannabis resin. The INCB had nothing to say about personal use though; it was primarily concerned with the international movement of drugs, which it wanted to regulate and stop if illicit.

Cannabis trafficking and regulation also came under the scrutiny of the UN's 1988 Convention against Illicit Traffic in Narcotic Drugs and Psychotropic Substances. Because the earlier agreements had not covered some of the more nefarious aspects of cannabis use in an international context, this convention expanded its purview to cover drug traffickers specifically and any precursors of the active chemicals in cannabis. Again, personal possession was avoided as an issue, which left it up to the various countries signing the convention to enforce their own laws.

Canada had its own particular legal problems with cannabis. A survey taken of Canadian citizens in 2018 revealed that a little over half had used cannabis illegally, while a little under a quarter were using it legally as a medicinal. In that country medical marijuana had been legal since 2001, and as the general attitudes of Canadians grew increasingly positive to cannabis, problems with the prosecution of marijuana users and buyers arose. Later in 2018 the Cannabis Act was passed, legalizing botanical marijuana use in Canada. Canada offers a notable example to other countries regarding the process of legalization at the national level and the factors involved in regulating the legal use of marijuana. Canada also produced an interesting conundrum for the US, because it borders some states with lax marijuana laws and other states with strict laws.

A single government such as Canada's can take steps toward legalization, and a uniform system can get established in that country. But drug policies in different nations diverge, and the lack of uniformity across borders can cause huge problems. As long as most nations held the same legal and social opinions on narcotic drugs, international regulation was simple; but when opinions change from country to country, consensus becomes elusive. Because of the extreme effects of opium use, international regulation of that drug could be considered relatively stable. But because of its proposed medical uses and the much milder impact of marijuana on society, attitudes about cannabis have diverged among countries. Collins suggests that for marijuana "a more flexible international system which both allows and provides a check upon divergent national cannabis policies is likely the best way to protect global health and welfare." The challenge is to move away from the hard-line international policies and on to a more flexible system. Brittney Griner's defense in Russia included the claim that she had been prescribed cannabis for pain caused by the many sports injuries she has incurred during her illustrious basketball career. Even though the marijuana she was caught with was considered medicinal in the US, because all categories of marijuana are absolutely illegal in Russia she was prosecuted and sent to a gulag. If the view of the international legality of marijuana had a more flexible basis, perhaps Griner would have been spared incarceration.

Moving Forward in the United States and Globally

Perhaps the biggest step in moving legalization forward in the US at the federal level is the rescheduling of marijuana in all forms from Schedule 1 to a different category. The Agriculture Improvement Act of 2018 went a long way to paving the way for federal legalization of cannabis by removing cannabis with less than 0.3 percent THC from Schedule 1. On the other hand, more concentrated marijuana products can only be optional for legalization if they are moved off the Schedule 1 list. Even with the Agriculture Improvement Act, there was much confusion when state and federal laws clashed. Legal expert Sam Kamin points to the *Big Sky Scientific LLC v. Idaho State Police* case in 2019 as a particularly sticky situation. This case involved the transport of 7,000 pounds of hemp from Oregon to a processing plant in Colorado. As both Oregon and Colorado recognize hemp (less than 0.3 percent THC concentrated marijuana) as legal, and the federal government is in agreement, this transport should have been perfectly fine.

The problem was that to get from Oregon to Colorado the driver of the transport truck traveled through Idaho, which does not view hemp as legal. Consequently, the Idaho State Police confiscated the shipment, prosecuted the driver of the truck, and announced that they had just made the largest drug bust in the state's history. They claimed that this arrest was legal under Idaho's anti-marijuana laws. Usually when a state law conflicts with a federal law, the state law gives way to the federal. Not so here, as the federal court adjudicating the case decided that Idaho had the right to seize and prosecute. The federal act should have trumped the Idaho law. But because of definitional confusion, the federal court undercut the federal law and sided with the state of Idaho. Hopefully such confusion will be cleared away as the Agriculture Improvement Act of 2018 gets recognized and interpreted. The *Big Sky Scientific* decision could be the result of the Agriculture Improvement Act being so new that it remained more logical to fall back to the state law.

Two other areas of persistent consternation concern ways that business is done in the marijuana industry and issues governing employment. Most banks are reluctant to take on marijuana industry accounts due to the Schedule 1 nature of many of their products and the current cash-only basis of most transactions. Even though the US Treasury Department has encouraged banks to engage in business with marijuana industry companies, many banks remain concerned that some form of money laundering might be involved. In late 2019 the US House of Representatives passed the SAFE Banking Act, which, as Kamin points out, "would have cleared the way for federally insured banks to offer services to the marijuana industry." But the Senate killed the bill, oddly because of a split among factions of the pro-marijuana legalization movement.

The Internal Revenue Code's section 280E hits marijuana businesses with a double whammy that makes the business fairly unlucrative. In essence this code makes marijuana businesses pay taxes on the product but does not allow deductions except on the cost of goods sold. Since the goods sold are only a small part of the business, this leads to industrial marijuana tax rates being as high as 80 percent of revenue.

Another court case that added more confusion to the fray was *Coats v. Dish Network*. In this case a worker for Dish Network, Brandon Coats, tested positive for marijuana and was fired in 2010. Coats was a medical marijuana patient with prescriptions that treated his muscular spasms. While medical marijuana was already legal in Colorado at the time, and the state also has a law prohibiting employers from firing someone for engaging in a legal off-duty activity, he was still fired. The Colorado Supreme Court ruled that because marijuana was illegal at the federal level, Dish Network had fired him legally. The problem was that the state law protecting employees from termination as a result of an off-duty legal activity was not specific enough. Rewriting the law or specifying more directly how the law should work would have negated the Coats case decision. Some states' marijuana laws are such that use of the drug is permitted in employment cases, but others allow employers the right to terminate employees with zero-tolerance policies.

Kamin recognizes that the legalization issues for marijuana are not simple. But one thing is obvious: there needs to be more uniformity to settle the issues. The federal level needs to get its act together, and then either the states will follow or their laws will be superseded by federal laws. Clarifying federal policy on cannabis use is crucial to moving forward with legalization. In 2020 Kamin recommended that "a Biden administration could look to the hard decisions being made in the states—regarding who may participate in the industry, how to make amends for the prior harms of the Drug War, and how to regulate marijuana in the public interest—to guide federal policy going forward." It remains to be seen how the second Trump term will affect the legalization issues.

The legal complexities regarding marijuana use reflect how changing mores and attitudes impact both cultures and political systems. While the US, Canada, Australia and a handful of other countries have taken legalization steps, it is remarkable that the rest of the world's nations have moved slowly or not at all. Attitudes toward cannabis continue to reflect highly divergent cultural values and expectations.

16

Dangerous?

In their classic book entitled *Marihuana, The Forbidden Medicine*, Lester Grinspoon and James Bakalar outline the many reasons why cannabis is an excellent candidate for the development of medicinals. They examine the medical utility of marijuana in the treatment of dozens of disorders. Each disorder that the pair believed was treatable with marijuana is discussed at length, with many "N-of-1" descriptions (taken from multiple crossover single-patient clinical trials). They also point out that the main reason for some of the US-funded work done on the safety of marijuana in the 1990s was that the US government was seeking good reasons to keep marijuana on the Schedule 1 list, hence funding studies investigating whether marijuana was harmful to users. Greenspoon and Bakalar took this notion of cannabis smoking as dangerous to task.

First, they argued, millions of people have smoked marijuana for thousands of years and there are no credible instances of lethality caused by marijuana; the number of nonadverse encounters with marijuana have far outweighed lethal or adverse ones. They then present a convincing table reproduced here (table 16.1) showing safety factor data for secobarbitol (a sedative), alcohol, and marijuana. The safety factor is a reliable measure derived by dividing the lethal dose of a drug by the therapeutic (also called effective) dose. They had one problem, however. There is no calculable lethal dose for marijuana, as there are no definitive data on

Table 16.1. SAFETY FACTORS FOR THREE SUBSTANCES

Drug	Effective dose	Lethal dose	Safety factor
Secobarbital	100–300 mg	1000–5000 mg	3–50
Alcohol	0.05–0.1%	0.4–0.5%	4–10
THC	50 mg/kg	2,160,000 mg/kg	40,000

Note: The alcohol data are blood alcohol level measures. The THC data are given as a function of the body weight of the organism being studied.

how many people have died from its use. So they used the lethal dose (LD) of marijuana in other animals (specifically mice), which makes their estimate fairly conservative. The LDs for mice, dogs, rats, and rhesus monkeys are known, and they are 2.2 g/kg body weight, 3 g/kg, 1g/kg, and 9 g/kg, respectively.

THC is about 10,000 times safer than either the sedative or alcohol. It takes a lot of THC to kill: 2.1 mg/kg is equivalent to two grams, which is about a half teaspoonful of THC (that's purified THC, not just ground marijuana). This brings us to "dabbing," the physical-chemical process by which THC/CBD content in a product can be increased to much higher concentrations.

Dabbing

The concentration of CBD in any particular marijuana product can vary considerably, except for those products labeled "hemp," which must have a THC content of less than 0.3 percent. But as many readers who have used marijuana will have noticed, the THC content of cannabis strains has increased dramatically over the past few decades. The grass available today is not what your parents encountered. A typical current strain will have up to 15 to 25 percent THC, which is quite concentrated compared with most strains from decades ago.

Dabbing involves increasing the concentration of THC to over 60 percent and even up to 90 percent by concentrating the THC into dabs, wax, or oil. In these forms the product can be vaped or smoked directly in a dab "rig" (a sort of bong pipe). Various names have been given to dab or concentrate that are quite descriptive of their consistency, including shatter (a glasslike extract that fragments upon blunt contact), crumble (a crumbly extract), budder (buttery consistency), live resin (made from frozen plant material and resins from most of the terpenes in the plant), distillate (distilled THC used to produce vape oils), terp sauce (runny, with a high terpene concentration), and crystals (powdery consistency from crystalized THC/CBD).

People in India have long made a familiar form of dab or concentrate—hashish or hash. The old-school way of preparing hashish (also known as charas) is to roll it until it forms a gooey ball. Cannabis flowers are obtained, bunched into a loose ball, and rolled or sifted or pressed. The amount of rolling depends on how meditative and persistent the person rolling the ball is. The rolling releases the resins and oils that were clinging to or part of the flowers. If one is persistent or sufficiently meditative, a dark resin will start to collect on the hands and the rolling surface. This resin is collected and rolled further into the distinctive spheroid objects that temple balls should look like. I have tried this; from one large flower of sinsemilla (a seedless strain), I was able to roll about a 5 mm diameter spheroid. I am sure I lost much of the starting material to inexperience, and it was not easy nor a zenlike occasion.

This rather primitive way of mechanically extracting the oils from cannabis can be improved upon by knowing some basic physical chemistry and a few details about the cannabis flower. All methods for making dab involve separating the trichomes of the flower from the rest of the plant. There are organic chemistry methods such as dissolving flowers in ethanol and then evaporating the ethanol away. Another approach involves drenching ground-up flowers in butane, which coprecipitates the THC, CBD, terpenes, and other chemicals from the flower. To get the unpleasant butane out, the lower boiling temperature of this chemical is used to distill it away, leaving the oils and extract. Another approach takes advantage of the chemical fact that cannabis oils are denser than water. If one freezes cannabis flowers and then rinses the frozen buds in cold water and collects the water, the water will contain a large quantity of the oils originally in the flower. Filtering the water completes the extraction process. Because dabbing uses oils with high concentrations of THC, one needs to be careful with dosing when using this kind of product.

The Dangerous Professor

There are many ways to determine whether a drug is dangerous or not, such as the safety factors mentioned previously. Another is to estimate its relative harm-

THE MOST DANGEROUS DRUGS

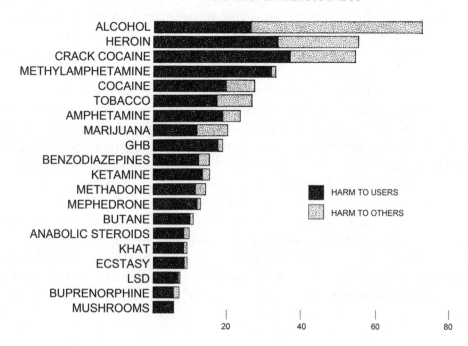

Figure 16.1. Relative danger of several drugs categorized as harm to self and harm to others. Adapted from Nutt, King, and Phillips (2010).

ful effects. This is what neuropsychopharmacologist David Nutt did in 2007 and 2010. He gathered a group of drug use policy experts, locked them in a room for a day, and had them rank the twenty drugs in figure 16.1 into sixteen categories. Roughly half the categories related to harm to oneself and the other half to harm to others. The results were illuminating, even though the study has been criticized for the past decade. The conclusion was that alcohol, heroin, and crack cocaine are three times more harmful than marijuana. But drugs such as LSD, ecstasy (methylenedioxymethamphetamine), and mushrooms were ranked as half to a third as harmful as marijuana. Moreover, while marijuana, alcohol, heroin and crack cocaine were considered roughly equally harmful to users as to others, LSD, ecstasy, and mushrooms were believed to pose minimal risk to others.

The problem with these studies is that each drug has a different set of reasons for why it can cause harm. This detracts from the computation of total harmfulness and overlooks more important issues that are involved in developing policies for drug use. In addition, the rankings were done by British medical and policy experts. The profiles would almost certainly be different in the United States or Canada. For example, while in England there is only slight harm to others caused by the use of methamphetamine, meth use in the United States would have a much higher ranking for harm.

Nutt is in fact a strong advocate for drug use. He lost his position as chair of the United Kingdom's Advisory Council on the Misuse of Drugs because of his suggestions that the government should reconsider its drug policies, and he was dubbed "the dangerous professor" for advocating more rational policies controlling narcotics. He believes the drugs at the bottom of figure 16.1 should be considered recreationally and medicinally safe. Much of his work after the 2010 study has focused on LSD and its potential utility in medicine. According to his rankings, this drug has one of the lowest self-harm estimates and causes little to no harm to others. His bottom line with respect to marijuana is that while it is harmful in some ways, it is less harmful than the highly addictive drugs at the top of figure 16.1. This lends some credence to Grinspoon and Bakalar's claim that the cost-benefit ratio of marijuana is so low as to beg for its use in medicine and other activities.

Physiological Damage to the Human Body

The main destination of THC in the human body is the brain and other parts of the nervous system. CBD has a higher affinity for other parts of the body. Hence the impact of CBD on the body will differ from the impacts of THC. The major reason that cannabis was placed in the Schedule 1 category is that many policymakers, scientists, and physicians believed that it was a harmful narcotic which required regulation.

There are two major ways that cannabis could be dangerous to our health. The first concerns the simple wear and tear the plant compounds might have on our physiology. The second involves the possibility that cannabis might be addictive. Both are indeed dangers that should be examined and, if real, balanced against the positive effects of cannabis. If the dangers far outweigh the benefits, then there is no reason to allow continued use of the plant or its compounds in medicinal or recreational situations. These issues are at stake in ongoing clinical trials, and the hard data collected in such trials can tell us much about the potential dangers or side effects. Some studies have already pointed toward potential damage to the body.

In 2018 public health specialists in Canada decided to dig down into the research to see whether already completed studies had clarified whether dangers existed for the use of cannabis. This kind of study is called a meta-review—a review of reviews, so to speak. Strict methods are used in such reviews to weed out irrelevant or compromised data. K. Ally Memedovich, Laura E. Dowsett, Eldon Spackman, Tom Noseworthy, and Fiona Clement collected sixty-eight studies of marijuana's impact on human bodies to carry out their analysis. They assigned a grade to each of the reviews based on criteria set by scholars who do systematic reviews. The scoring rules are exacting, and to survive with a high score (9 or more out of 11) is no simple trick. While all the studies were included in the overall summary, the high-quality scores enabled the researchers to assess the relevance and believability of the results. Their analysis revealed that only about 15 percent (eleven out of sixty-eight) of the studies included in the review were of high quality. Given the overall low quality of the studies in the meta-analysis, what can we extract from them? If, for instance, the data are scattered randomly, then one reason might be the studies' mediocre quality. The major trends highlighted by the analysis, however, are quite intriguing and potentially important.

The overall analysis indicated that there are indeed health issues that need to be considered when using marijuana either medicinally or recreationally. Sixty-eight physiological, neural, and behavioral aspects were examined, of which sixty-two showed detrimental outcomes. A large number of these harmful outcomes concerned the impact cannabis has on the brain and specifically on mental health, including psychosis, mania, and suicide. Figure 16.2 summarizes the results of this seminal study.

Cannabis use also affects organ systems other than the brain. Memedovich and colleagues point out that their study indicates "association with stroke, atrial fibrillation, bronchodilation, respiratory complications and chronic obstructive pulmonary disease." There also appear to be significant harmful impacts on some cancers. One of the more surprising harms not examined by Memedovich and colleagues is an impact on spermatogenesis in the male testicle, demonstrated in some nonhuman animals. Studies on dogs have shown that sperm synthesis in the testicles of dogs treated with cannabis (12.5 mg/body weight) is inhibited for about thirty days after the treatment. The same results were observed for mice, where 2mg/day THC doses were administered for forty-five days. The production

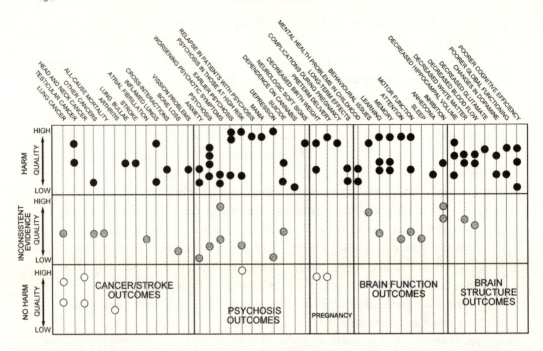

Figure 16.2. Result of the Memedovich et al. (2018) meta-analysis (systematic review), showing the assessment of forty-four neural, physiological, and behavioral outcomes. There are ninety-three dots in the figure representing the sixty-eight studies included in the review. The reason for more dots than studies is that some studies assessed multiple outcomes. Note the low number of "no harm" and "inconsistent evidence" outcomes. There are sixty-four dots in the "harm" category.

of sperm was completely stopped after such treatment. Luckily for the mice and dogs so treated, the effects were reversible by abstaining from ingesting THC. This could be a serious consideration if you intend to reproduce and smoke marijuana at the same time. Other effects on our physiology include irregular heart rates, blood pressure fluctuations, and dry mouth.

From a purely physiological/neurological perspective, drugs don't belong in our bodies. So why do we ingest compounds such as alcohol? My colleague Ian Tattersall and I have elsewhere discussed the reasons we drink alcohol, and many of these reasons also apply to drugs such as cocaine, heroin, marijuana, and hundreds of other compounds that humans have ingested at one time or another. Much of our predilection for ingesting these compounds concerns our desire and sometimes need to alter our consciousness. But in altering consciousness, drugs come into contact with our nervous system and our brains in particular. It is no wonder that drugs would have some adverse effect on these tissues. Memedovich and colleagues noted both physical changes to the brain and behavioral changes induced by cannabis use.

Several studies have assessed structural and chemical changes in the brain caused by marijuana use. In the structural change studies, high-tech brain imaging methods such as magnetic resonance imaging (MRI) have been used to compare normal brains and "marijuana brains." These studies show some subtle changes in the volumes or sizes of the amygdala, hippocampus, and white/gray matter of the human brain. Volume changes of brain structures can be indicative of change in function. There was a famous study of London taxi drivers, who must commit to memory the names and locations of more than 10,000 streets in that city to obtain a license. The part of the brain responsible for memory (called the hippocampus) was enlarged in these taxi drivers, presumably as a result of such memorization.

Getting larger is usually better than shrinking in these kinds of studies, and since these particular brain structures often seem to decrease in volume the inference is that it's not a good outcome. That's the bad news; the good news is that overall brain volume, intracranial volume, and the corpus callosum are unchanged in "marijuana brains." At the chemical level, several neurotransmitter levels show disruption from normal concentrations. These transmitters include glutamate, dopamine, and choline, and their altered concentrations must surely impact brain function and how a marijuana brain responds to the outside world. Indeed, decreases in learning, memory, and attention span are some of the major shortcomings of a marijuana brain.

The marijuana brain is particularly relevant to the development of the adolescent brain. The connectivity of neurons that become essential for cognition later in life is an important developmental outcome of the teenage brain's transition into an adult brain. Some evidence suggests that long-term effects may persist in the brains of teenagers who use marijuana. With some variance, the age of sixteen appears to be a critical age for brain development, and this is why understanding marijuana usage in eighth to twelfth graders in the United States has become a priority. High school users who start smoking marijuana before the age of sixteen fare poorly on many kinds of tests relative to nonusers. Planning, decision-making, and memory are compromised by marijuana use before the age of sixteen, suggesting that cannabis affects neural connectivity in the developing brain.

Because marijuana changes the brain in the way it does, researchers have examined the impact of the plant on certain aspects of our behavior and mental health. Perhaps the major interest has been the possible role marijuana might play in addiction. Memedovich and colleagues' meta-review focused on studies of cannabis's impact on psychosis/schizophrenia, anxiety, suicidal tendencies, depression, and mania. Figure 16.2 shows that all these mental health outcomes are affected by cannabis use. Specifically, increased risk of symptoms of schizophrenia or psychosis was linked to heavy use and even average use of cannabis, compared with never using it. Early onset of symptoms of psychosis was also related to cannabis use. These data indicate an alarming relationship of cannabis use to triggering mental illness in those predisposed to the disorders. Equally alarming is the association of cannabis use with several behavioral precursors to suicide or mania.

These potential harms cannot be easily dismissed. They need to be considered if cannabis use is recommended by a medical professional or when a naive user begins to experiment with cannabis as a recreational. Just as one wouldn't feed peanuts to someone with a peanut allergy, so we should be alert to the potential for adverse effects from using cannabis on a regular basis.

Is Addiction to Cannabis Real?

Two of the major arguments levied against marijuana use during the time of Anslinger's Army in the 1930s and 1940s was that it was highly addictive and led to harder drug use (the gateway hypothesis). Contrast this with the 1944 La Guardia report, which posited that marijuana does not lead to addiction in the medical sense of the word and does not lead to heroin or cocaine use. The controversy over these issues hasn't gone away. Let's first look at how addiction is defined and then at the biological/psychological addictive properties of two drugs where addiction is a severe problem—cocaine and heroin—to explore what truly addictive drugs do to our brains and bodies.

Addiction is usually defined as a strong inclination to do, use, or indulge in something repeatedly, despite the dangers involved. In this definition, anything with a dangerous outcome can be addictive. "Danger" is the key word here. Any addictive behavior can become dangerous because of the eventual reliance a person places on the addictive substance or act. Gambling, sex, and some other behaviors are dangerous because of the sociological or financial problems they create, and hence can be considered highly addictive. Substance dependency is probably the most common kind of dependency.

To understand the chemical basis of addiction, we have to turn back the evolutionary clock to the development of our mammalian brain. Our complicated brains evolved to get the rest of our bodies to seek rewards. Whether it be for food, drink, sex, or a high, our brain chemistry evolved to focus on a reward system that when stimulated strives for more of the stimulant. What is the reward? In chemical terms it is basically the molecule dopamine, which when bound to dopamine receptors in brain cells stimulates an elated and euphoric feeling. If this euphoric feeling persists as a result of continued dopamine presence, then we seek more, although there are molecular mechanisms in the synapse called reuptake mechanisms that will clear the dopamine. Addiction happens when some stimulant we obtain from outside of our bodies tricks our reward system by allowing dopamine to linger in our synapses.

The primary recreational neurological mechanism for cocaine's use is to bind to dopamine receptors and produce that euphoric feeling. A normal dopamine reaction in the brain's reward center is followed by the clearing of dopamine through the reuptake process, where the dopamine is moved from the synapse and sequestered back inside the presynaptic neural cell. The problem with cocaine is that it doesn't allow the reuptake of dopamine. Instead it hangs out longer in the synapse, producing a craving for more and more. It takes a lot to clear the synapses of this effect, leading to a psychological and often physical dependence on the drug.

For an opioid the primary medicinal neurological mechanism is an analgesic one. For recreational use its primary mechanism is the same, but it can create a euphoric state for the user. Prescribed opioids are administered to stave off pain, and they do so effectively by binding to opioid receptors specifically in the midbrain. Once they bind, pain (which is perceived by this part of the brain) is alleviated. This binding involves blocking the neurotransmitter gamma-aminobutyric acid (GABA). By affecting GABA in the midbrain, opioids also access the reward system part of the brain and interfere with the normal interaction of GABA with dopamine receptors. This interference leads to dopamine overload and to addiction.

Both kinds of addiction (cocaine and opioids) effectively hijack the reward system, creating a terrible feedback loop that is often inescapable and leads to physiological and neurological dependence on the drug. Now let's consider marijuana and its impact on the brain, so that we can critically assess the addictive action of cannabis. Remember that THC interferes with GABA regulation, and this in turn interferes with dopamine action. If all that marijuana had in its chemical armory was CBD and no THC, any addictive properties of the plant would probably be minimal, as CBD does not interfere with the dopamine system. So it is the action of THC that we need to assess when considering marijuana addiction; indeed, any abuse-related outcomes of marijuana use are controlled by THC.

Cannabinoid receptors 1 (CB1) are part of what is called a "retrograde" signaling system. Moreover, the primary targets of THC in our bodies are the CB1 receptors, concentrated to a degree in our brains. The natural binding of the endocannabinoids anandamide and 2-AG to these receptors constitute the first steps of the endocannabinoid system. But as with any neurotransmitter system, once the binding occurs, the neural cells must next clear those transmitters from the synapse to prevent a response overload of those neural cells. The clearing mechanism for the endocannabinoid system is accomplished by enzymes called hydrolases and lipases. Also recall from chapters 10 and 11 that the endocannabinoid system interacts with the GABA and glutamate neurotransmitter systems. This means that the reward system then gets involved. One theory of THC addiction is that when THC binds to a CB1 receptor and activates it, there is a concomitant inhibition of GABA release, which enables the dopamine neurons to fire. The amount of firing regulates the degree of the impact of dopamine and in turn the degree of euphoria. This hijacking of the GABA system leads to the same problems that interference with the GABA system in alcohol addiction and opioid addiction do, although the level of euphoria and strength of the interference for THC is closer to alcohol than opioids. Given the way that THC interacts with the GABA system, there is a distinct potential for addiction for this compound.

Cannabis Use Disorder

A primer on cannabis addiction by researchers Jason P. Connor, Daniel Stjepanović, Bernard Le Foll, Eva Hoch, Alan J. Budney, and Wayne D. Hall points out that cannabis dependency is best broadly defined as the inability to stop consuming cannabis even when it is causing physical or psychological harm. In their terminology this is called cannabis use disorder (CUD). Connor and colleagues observe that "in 2018, the United Nations estimated that 192 million persons or 3.9% of the global adult population had used cannabis in the previous year. . . .

Approximately 9.9% of individuals who reported cannabis use in the past year were daily or near-daily users. . . . According to the most recent global estimate 22.1 million persons met diagnostic criteria for Cannabis Use Disorder in 2016 (289.7 cases per 100,000 people)."

Cannabis use disorder has become a major subject for study among medical professionals and human geneticists. Over the span of one recent year (2021–2022), at least twenty publications examined the genetic data on cannabis use disorder, many trying to pinpoint any human genes that might be associated with cannabis dependency. These publications generally shared the same conclusion: there is indeed a genetic basis for cannabis use disorder. The major method they adopted was the GWAS approach described in chapter 13, which uses the variation of the population being examined and a case-and-control experimental design to pinpoint regions of the genome that are associated with the cases. In the case of cannabis use disorder, entire genome sequences for large numbers of humans and the metadata on individuals' usage patterns were employed.

Prior to 2020 several GWAS studies were performed on cannabis use disorder with tantalizing results but also with caveats. One of the bad raps on these earlier studies was the sample sizes, which used slightly over 2,000 cases. For GWAS, increasing sample size often increases the accuracy of the inferences made. In 2020 a publication with over 200 authors presented results of an extremely large GWAS study of cannabis use disorder and related disorders. These researchers used data from 363,884 controls and 43,380 cases. Previous to this study, the sample sizes were only about a tenth of this size.

Many human traits have highly complex genetic architectures. For instance, height is apparently controlled by hundreds of genes in our genome. Nearly 300 genetic loci have been tabbed as associated with schizophrenia (which is probably several different disorders). One might expect cannabis use disorder to be rather complex too. Surprisingly, though, the cannabis use disorder GWAS revealed fewer than ten associated genetic loci. Of these only two passed muster when more stringent criteria were used to identify the best associations. These two loci are CHRNA2 (as well as a gene called EPHX2, which is linked to or in very close proximity to CHRNA2 in the genome) and FOXP2. The former (CHRNA2) showed a strong association with CUD in previous studies and has a cellular role as a gated ion channel protein. Its activity has also been strongly implicated in nicotine abuse. The connection of CHRNA2 to cannabis use disorder is tantalizing, since both nicotine abuse and cannabis use disorder involve ingestion by smoking. Equally tantalizing is that CHRNA2 has also been shown to be one of the many genes associated with different forms of schizophrenia. As noted earlier, schizophrenia in those predisposed to the disorder can be triggered by marijuana ingestion.

The FOXP2 gene codes for a protein that is a transcription factor. It is one of those genes that is involved in many human traits. It appears that it is important in regulating the expression of many different genes. Surprisingly, it has been associated with other behavioral traits such as age at first sexual intercourse, generalized risk tolerance, and educational attainment. Some of the cellular roles of FOXP2 also point to a neurological role, as it has been shown to be essential to synaptic plasticity. Its ubiquity in gene regulation makes it difficult to be more precise about its role at the cellular level in cannabis use disorder.

While the GWAS studies are intriguing and point to interesting genes, pinpointing exact causal roles is still beyond the reach of researchers. But there are potential remedies for cannabis use disorder. The first and most obvious is abstinence, but relapse into cannabis use disorder is common among those with the disorder when they find themselves in the presence of the substance or when in stressful situations. Cannabis use disorder involves many regions of the brain (CB1 receptors are particularly plentiful in the brain), and so targeting a specific brain region for treatment is tricky. Cannabis use disorder withdrawal symptoms can be treated with drugs such as dronabinol, nabilone or nabiximols. Finally, the pathway to anandamide synthesis can offer treatments. The idea here is to displace THC from the neural system with the natural endogenous neurotransmitter anandamide. Increased anandamide can reset neurotransmission patterns and alleviate the symptoms of withdrawal.

Overuse of any psychoactive substance such as alcohol or marijuana is something to be aware and wary of. As toxicologist Andrew Monte puts it, "Cannabis is not the root of all evil, nor is it the cure for all diseases. . . . You've got to understand what the good is and what the bad is, and then make a balanced decision." Grinspoon and Bakalar were correct in pointing out that cannabis offers a treasure trove of medicinals. But any drug or treatment needs to be balanced with its side effects. In many ways modern marijuana use has sidestepped a lot of the reckoning concerning the side effects. But legalizing it has brought cannabis into the limelight, where it can now be scrutinized and understood better. Balance is everything.

17

Contemporary Cannabis Culture
in Nine Graphs

Marijuana has been infused into our contemporary culture in such an extreme way that this book can only minimally touch on its current effects. Because our future use of the plant and its compounds will be affected by changing social, political, cultural, and philosophical attitudes, we need to consider some of the trends that cannabis culture has experienced in the past few decades and might in the future. I will consider the trends in attitudes toward legalization, marijuana potency, recreational and medicinal use in the United States, stock indices, spending on US research, use by age, global distribution of cannabis, and cannabis's influence on our entertainment.

Legalization

The most recognizable changes in attitudes toward marijuana followed its legalization in California in 1996. Figure 17.1 summarizes the results of Pew Research Center polls taken from 1969 to 2020. I show only the years 1996 to 2020; this graph depicts the overall trend of increasing support for legalization (white for

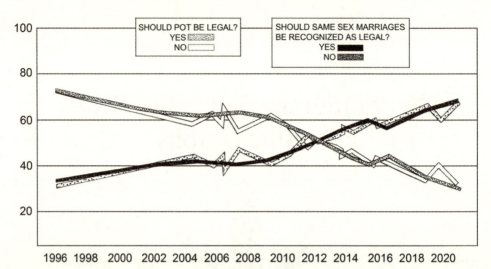

Figure 17.1. Survey results from 1996 to 2020. The first survey question was, "Should pot be legal?" White = no, light gray = yes. The second survey question was, "Should same-sex marriages be recognized as legal?" Dark gray = no, black = yes. Marijuana poll from Pew Research Center; same-sex marriage poll from Gallup.

legalization and dark gray against it). The other lines in the graph (black and light gray) are data from Gallup polls taken from 1996 to 2020 involving people's attitudes to gay marriage. The lines overlap considerably, with the crossover point coinciding nearly perfectly in the two surveys. Are attitudes to legal gay marriage driving attitudes to cannabis legalization or vice versa, or is there no relationship at all?

The key word for understanding graphs such as this one is "trend." Its definition is "a general direction in which something is developing or changing." Epidemiologists use it to summarize the changing dynamics of a disease that spreads in populations. Interpretations of a trend can serve as a basis for hypothesis testing and future research, and can suggest ways to deal with the overall trajectory of a phenomenon. A trend in itself won't test a hypothesis, though. It can only suggest the presence of something, and further study is necessary to determine the causation of the phenomenon. The first lesson from this cannabis graph is that it is a simple representation of real data. These kinds of graphs are persuasive because they are based on real data, as opposed to graphs that simulate outcomes.

The second observation is that there are four lines and hence four trajectories. A trajectory is not a trend, but rather a visual representation of the course of a measured variable over age or time that leads one to see a trend. Assessing the trajectories of different categories is an important step toward perceiving a trend and understanding its nuances. Once the trajectories are plotted, a trend might or might not be evident. If there is no trend, that's not so bad, because the analysis tells you something important: there are no changes in the Y variable with time, and that could be critical information. But fig. 17.1 shows a continuous trend to more acceptance of legal pot over time. Likewise for attitudes to gay marriage: there is a continuous and positive attitude to legal recognition of same-sex marriage.

The trends in the graph can now be stated as, "The attitude toward legalizing pot changed from mostly negative to mostly positive from 1996 to 2020. Likewise, the attitude toward the legal aspects of gay marriage changed from mostly negative to mostly positive from 1996 to 2020." Even with these more clearly stated trends, we still cannot say anything about gay marriage attitudes having a causal relationship with attitudes to legalizing pot. We can only say that it appears that trends in positive response to legal pot are correlated with the attitudes of people toward gay marriage. Gay marriage may not (and most probably does not) have anything at all to do with pot smoking; we can't say anything about causation.

How would we go about demonstrating, for example, that attitudes to pot *caused* the trends of attitudes to gay marriage (or vice versa) in the graph? This endeavor would involve posing a hypothesis and testing it. The hypothesis would be something like the null hypothesis (typically the null hypothesis is written as H_0) = Attitudes to legalization of pot are directly responsible for increases in positive attitudes to gay marriage. An adequate test of this hypothesis would no doubt lead to its rejection. But the way science works is that rejected hypotheses can help advance us. When a hypothesis is rejected, we move on to a more restrictive or specialized hypothesis that can be tested to find an explanation for the correlation.

In subsequent studies one might interview people with a cleverly designed survey that strives to understand the sociological reasons why, at a particular point in time in the 2010s, people's attitudes to pot and gay marriage reversed. Maybe politics was involved; maybe religion; maybe it was Barack Obama's presidency; or maybe it was something else a creative social scientist can pinpoint. But any causation will remain a mystery until a hypothesis is tested. Using the scientific method, we can whittle down the possibilities and find better explanations for phenomena we see in both nature and culture.

Is Cannabis Becoming More Potent?

It is commonly believed that cannabis products are getting stronger and stronger as time goes on. What can we discover about that perception? Fortunately, agencies such as the FDA and DEA keep track of THC and CBD concentrations in confiscated marijuana. Figure 17.2 presents data from 1995 to 2018 compiled by the DEA. Specialized equipment (gas chromatography and mass spectrometry) is needed to accurately determine the THCA content of a sample, and these determinations are often highly accurate, but sometimes not. Nick Jikomes and Michael Zoorob have compiled data from over 300,000 analytical tests in Washington State to gauge the accuracy of THC and CBD concentration estimations. Their results "documented systematic differences in the cannabinoid content reported by different laboratories, relative stability in cannabinoid levels of commercial flower and concentrates over time, and differences between popular commercial strains." In general, THC levels can be illuminating, and there is no reason to doubt the overall trends demonstrated by figure 17.2. The results come from a single lab with the same standards (DEA). Measurements at different time points were taken, and only well-defined commercial strains were used.

The results demonstrated a fourfold increase in THC concentration from 1996 to present. The increase was somewhat linear over the three decades of the surveys. On the other hand, CBD concentration appears to have dipped over the years, even though the concentration measures in 1995 and 2018 were nearly the same. CBD concentration doubled between 1995 and 2002, but consistently dropped to a low 0.14 percent in 2017.

This trend for increase in potency is not just a US phenomenon. Seizure of illicit European marijuana and subsequent THC analysis have shown the same overall trend of increased potency in Europe.

My marijuana was nothing like my daughter's, who was born in 1996. What does this increase in potency indicate? One possible reason for the sharp increase is demand for a more potent psychoactive product. For a long time, there has been no regulation of THC content in illicitly grown marijuana, and growers could amp it up as much as possible. For recreational reasons (a better, more powerful high) growers did the breeding that increased the THC content. Another likely factor could be economics. According to pharmacologist Mahmoud A. ElSohly, "the higher the THC content is, the more expensive the product." Demand for more expensive pot was strong enough to induce growers to increase the THC in their

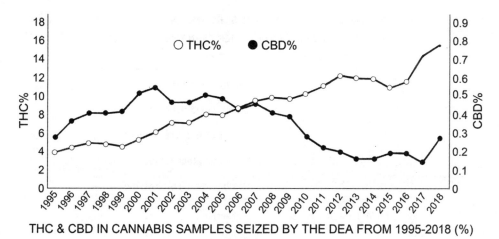

THC & CBD IN CANNABIS SAMPLES SEIZED BY THE DEA FROM 1995-2018 (%)

Figure 17.2. Proportions of THC and CBD content over the years since legalization began. Adapted from Chandra et al. (2019).

strains. This purely economic reason brings us to our next set of graphs addressing the economics of marijuana.

Marijuana as a Booming Business

There are many indicators for a healthy business. The first we examine here for the cannabis industry concerns patents. This indicator is similar yet distinctly different from the clinical trial indicators discussed in chapter 14. Patent analysts Joseph Wyse and Gilad Luria have provided an extensive review of cannabis patents as of 2021. They suggest that product development in the cannabis industry has three major areas: upstream agritech, midstream chemistry/analytics, and downstream medical applications (table 17.1). Each of these areas produces different challenges for patent development of medical cannabis products.

Patenting cultivars is another question. To date about 300 different cannabis strains have been patented, which means over 300 cultivars have been deemed distinct enough to warrant intellectual property rights protection. We can expect a deluge of patent applications as the legalization ramifications have more effect on cannabis farming and breeding. In addition, experimental strains produced by gene editing might also swell the number of patent applications in the cannabis

Table 17.1. UPSTREAM, MIDSTREAM, AND
DOWNSTREAM DEVELOPMENTS IN THE
MEDICAL CANNABIS INDUSTRY

Upstream	Improved strains
	Genetic modification
	Plant material assessment
	Harvesting technology
	Postharvesting processing
Midstream	Extractions
	Purification methods
	Separation methods
Downstream	Disease treatments
	Medical devices
	Compositions
	Formulations
	Dosage forms

industry. Wyse and Luria show that there was a significant peak for 2018 in the number of patent filings for all three stages of cannabis product development. They reasonably conclude that patent filing growth is related to legalization and the economic lure of the cannabis industry. The economic siren is what this section's graphs are all about.

It is undeniable that marijuana has become a booming business since its legalization in many US states and in other countries. The signs of its health as a business included increased sales and employment. Figure 17.3 summarizes global medicinal sales, global adult use sales, and job availability in the marijuana industry in the United States. The trend lines are all increasing, with jobs going from under 100,000 in 2017 to almost 500,000 in 2023. Simulations of job availability in 2024 and 2025 project another 100,000 or so workers in the cannabis arena.

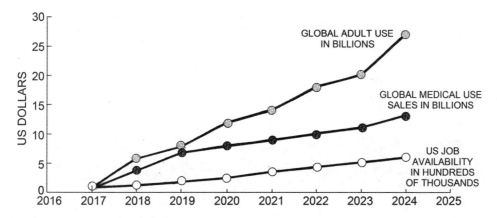

Figure 17.3. Summary of trends from 2017 to 2022 and simulated to 2024. The top line represents global adult use in billions of US dollars; the middle line represents global medical use in billions of US dollars; the bottom line shows the number of US jobs in the cannabis sector in hundreds of thousands. Adapted from ElSohly et al. (2021).

Sales are also trending upward, showing a tenfold increase in sales of medicinal marijuana and a twenty-five-fold increase in sales for adult use.

Another indicator of business health is the number of jobs added each year. Figure 17.3 shows the increase in jobs from 2017 to 2021 in the United States was about 100,000 new openings. The state-by-state increases are equally interesting— Florida led the way with over 9,000 new jobs, followed by Nevada, Washington, Arizona, and Colorado, all of which had between 4,000 and 7,000 jobs open. Pennsylvania, New York, New Jersey, Illinois, Alaska, Maryland, and Oklahoma were all doing quite well too, with between 1,500 and 3,000 new jobs created in the cannabis industry per state. Because each state has different population sizes, a better indicator of what these numbers mean would be the percentage of job gains in each state. This metric suggests that Pennsylvania (with a 4,200 percent increase in jobs) and Florida with (a 700 percent increase in cannabis jobs) are experiencing especially rapid growth of the cannabis industry work force. Another aspect of economic health concerns how much tax revenue governments take in from their regulation efforts. This variable is also trending upward.

Since legalization, the US stock market has seen many cannabis-based companies become publicly available. There are over 5,000 indices characterizing the stock market, the most popular of which are the S&P 500, Dow Jones Industrial

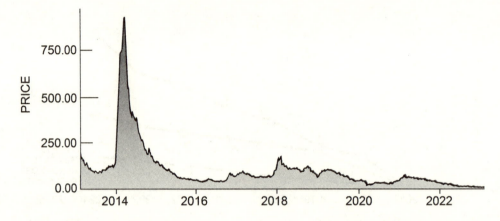

Figure 17.4. Price fluctuations in the Global Cannabis Stock Index from 2014 to 2023. Data from the Global Cannabis Stock Index.

Average, and Nasdaq Composite. For cannabis there are about ten indices that can be used to compare and contrast stock performance. New Cannabis Ventures (see For Further Reading) lists four indices and provides plots for stock performance by day and compressible options to view trends over ten years. Figure 17.4 shows a ten-year projection of the Global Cannabis Stock Index, based on the performance of publicly available cannabis entities. Other indices based on different assumptions and criteria will show different patterns, but the graph in figure 17.4 can suggest how the stock market influences our attitudes to cannabis and vice versa. One should always inspect the scales on the y-axis of a graph. What might look like a smooth period of stock performance between 2018 and 2023 in figure 17.4 (albeit a little jagged) can on a different scale appear wildly unstable. The caveat is that the timeframe examined here is over a decade, and the price of stocks ranged from $1,000 to a little under $10. There was a huge peak in 2014, when stock prices started at about a little under $100 and skyrocketed to $1,000. These prices persisted for about a year and then dropped precipitously to about $50 by 2016. This pattern is what stock market experts would call volatile.

Over the last eight years, though, the average stock price has stayed somewhat stable but bounced between a little under $200 down to about $10. Again, scale is important here. What I am calling a relatively stable period might look to others like a fairly volatile period. If stock was bought in 2016 and sold in 2018 a trader might make as much on average as $100 per share sold. On the day I sat down to write this chapter the Global Cannabis Stock Index was down to ten dollars per share. Other indexes had a bit higher average stock prices but in general as the figure shows there is a slow gradual decrease in stock price since 2021.

The dips and peaks can be correlated with specific events that impacted the value of cannabis stock. For instance, that blip up in 2018 coincides with the legalization of recreational cannabis in California (January 1, 2018) and also caused by legalization in Canada a few months later (October 17, 2018). The dip around 2020 is more than likely the result of California increasing its tax markup on cannabis from 60 percent to 80 percent. Following the stock market is tricky business, but astute knowledge of what is going on in the cannabis world (like legalization events and tax increases) can help in navigating the market. Being aware of how resources are allocated in marijuana product development is another way to keep up with the often times blurry trends of the stock market. We turn to these issues in the next graph.

Allocation of Research Money in Developing Cannabis as a Medicinal

Chapter 14 discussed the complexities of the drug approval process in the US. Part of drug approval has involved deregulating cannabis as a dangerous drug. Is marijuana being treated in a balanced way by government regulators? One way to address this question is to ask what the federal and state governments focus on when funding marijuana research. It would seem logical that if governments are supporting medicinal research and little else, one could conclude that the goal of regulation is to focus on expanding the medical applications of cannabis. On the other hand, if the major recipients of government funds are for research into the legal aspects of cannabis, then a different agenda could be perceived.

Figure 17.5 shows the trends for funding of five categories of marijuana research focus in the United States. Twenty years ago, the major focus of US research funding was for projects examining the harmful effects of cannabis use. A good portion of the research funding also went to studying the endocannabinoid system. Cannabinoid (synthetics) and *Cannabis* (the plant) composed a mere 10 percent of the total funding for cannabis research. As the graph shows, there have been steady increases in funding for all five categories of cannabis research.

But the focus on cannabinoids and cannabis as therapies hasn't gained much

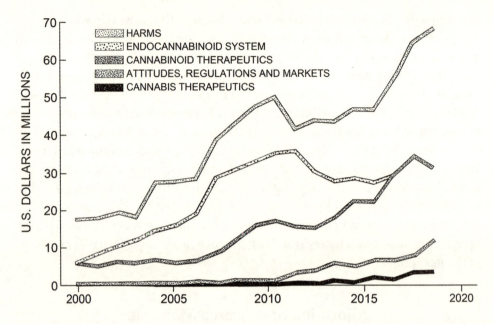

Figure 17.5. Graph of federal spending patterns on marijuana research from 2000 to 2020. Adapted from O'Grady (2020).

ground relative to studies investigating the harmfulness of cannabis and how the endocannabinoid system works. The amount of research money dedicated to therapeutics is only about 20 percent of the total research budget. Even social studies of the attitudes, economics, and regulations of cannabis are funded at a higher rate than cannabis therapeutics. Knowing the harmfulness of cannabis and cannabinoid impacts on humans is of course an important part of understanding cannabis's therapeutic potential, but it seems unfortunate that relatively little funding supports research on the therapeutic nature of cannabis and cannabinoids. Focusing on more specific aspects of the harms produced in a sociological context would also provide a better balance of research investment. We turn to this topic with the next graph, which examines use of cannabis by teens.

Are Young Children Smoking a Lot of Pot?

Oddly, high school kids have been placed in the midst of marijuana culture for much of the last century. From Ralph Wiley ("Faster, faster. Play it faster") in the movie *Reefer Madness* to Jeff Spicoli in *Fast Times at Ridgemont High* ("All I need are some tasty waves, a cool buzz, and I'm fine") to the offspring of the *That '70s Show* stoners in *That '90s Show* (who are themselves stoners), film has portrayed many American youngsters as pot smokers. And more than likely the biggest celebration of cannabis use today, 4/20, had its beginning in a high school. While there are many origin stories for the 4/20 celebration (for instance, the Bob Dylan factoid that 12 times 35—as in "Rainy Day Women"—equals 420), the most plausible story suggests that a San Rafael, California, high school was the wellspring. The kids at this high school would gather at 4:20 pm after classes to smoke marijuana. They would use "420" as a code for their sessions. The time 4:20 morphed into the date 4/20, which became a national marijuana day.

How prevalent is teen cannabis use in actuality? Under no current laws in the United States can marijuana be sold to children or young teenagers. It is unreasonable to assume, however, that they cannot get their hands on it. In places such as New York City, where open marijuana smoking is tolerated, bus stop cannabis-smoking sessions are common among high schoolers. In chapter 16 I mentioned the harm that can be done to the developing teen brain if marijuana abuse is part of a teenager's everyday life. But to drive home the dangers, I provide a list of hazards that marijuana use poses particularly for teenagers. This list comes from the US Substance Abuse and Mental Health Services Administration:

- Neuropsychological and neurodevelopmental decline
- Poor school performance
- Increased school drop-out rates
- Increased risk for psychotic disorders in adulthood
- Increased risk for later depression
- Suicidal ideation or behavior

Figure 17.6 plots marijuana usage patterns for eighth, tenth, and twelfth graders from 1975 to 2023. There are several interesting trends. First, although there are no data for 1975 to 1990 for eighth and tenth grades, the usage of teens starting in the 1970s rose to its peak around 1979 of about 50 percent of the respondents smoking marijuana regularly. Though illegal during this period, there was a great deal of interest in marijuana use by teens that gradually decreased to a low for twelfth graders in 1991 of about 20 percent. Second, the years after tenth and eighth graders were placed into the survey showed a gradual increase in usage from the eighth to tenth to twelfth grades. It is notable that all grade categories

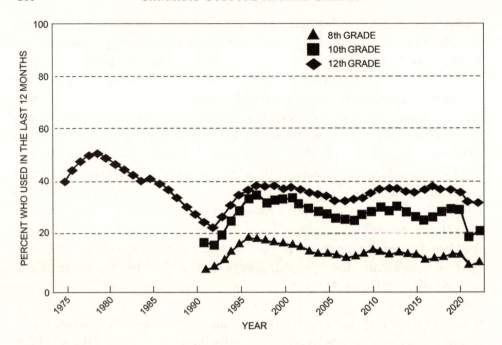

Figure 17.6. Trends in marijuana use among eighth-, tenth-, and twelfth-grade US students. Data from Monitoring the Future (https://monitoringthefuture.org/, also known as the National High School Senior Survey). Over 32,000 students from over 300 high schools were surveyed in 2020–2021. This ongoing study has been conducted by researchers at the University of Michigan.

showed a sharp increase in the 1990s, likely due to legalization/decriminalization in some states. Third, essentially the trends for eighth, tenth, and twelfth grades mirrored each other from 1990 to 2020; but eighth grade was always lower than tenth grade, which was always lower than twelfth grade. Fourth, there has been a stable pattern and maybe a slight decrease in marijuana use by teens in the survey since 1990 until 2020–2021, where a very noticeable dip in usage appeared at all three age levels. The dip appears to have stabilized, because the results for 2021–2022 were all upward relative to the previous year.

This fourth trend was stunning and not evident for other drugs that the survey considers. Even in the face of many states relaxing their marijuana laws, this considerable dip was observed. According to Nora Volkow, the director of the National Institute on Drug Abuse, which funded the survey, "We have never seen such dramatic decreases in drug use among teens in just a one-year period. These data are unprecedented and highlight one unexpected potential consequence of the COVID-19 pandemic, which caused seismic shifts in the day-to-day lives of adolescents." Volkow observed that, if researchers can pinpoint what made teens decrease their usage during 2020–2021 (beyond the fact of the pandemic), then perhaps we will have clues for reducing marijuana use in teens in the future. Perhaps the pandemic induced more parental oversight, or caused a lack of drug availability; perhaps other unforeseen social aspects were factors. At this point we don't know what exactly caused the dip, but it most probably has something to do with how our lives changed during the pandemic.

How Global Is Cannabis?

Chapter 15 reviewed the worldwide legality issues of cannabis. But what about actual global use patterns? Figure 17.7 graphically represents the global use patterns of cannabis. The two countries with the largest populations are not quantifiable, as data are lacking for India and China. The pattern of use is somewhat clear for the rest of the globe. North Americans are relatively heavy users of cannabis compared with the rest of the world, as are Australians and New Zealanders. European use is also quite high, with France topping the countries there at an overall use percentage of 11 percent. Some African countries such as Nigeria and Zambia also have heavy usage. Not surprisingly, the first country to completely legalize marijuana, Uruguay, has one of the highest usage rates. These results seem to be solid except for one glaring problem. Oddly, Iceland tops the chart with a whopping 18 percent usage rate. Marijuana use was at one time legal in Iceland but was made illegal in 1969. And alcohol was prohibited in Iceland up until about thirty years ago. Icelanders argue that the figure of 18 percent is ridiculously high. And it might be.

The study summarized in figure 17.7 used the percentage of the entire population that had smoked marijuana in the last year (which would be 2014 for the Iceland data). But the 18 percent figure for Iceland resulted from asking those who used marijuana if they had used it in the last year. This way of asking the question severely biased the overall result. If the survey had been done correctly, the total would show only 6.6 percent of Icelanders smoking pot in the last year, pulling it down to twenty-ninth on the pot-using list. Icelanders, although proud of their progressive attitudes, pointed out that marijuana was illegal when the

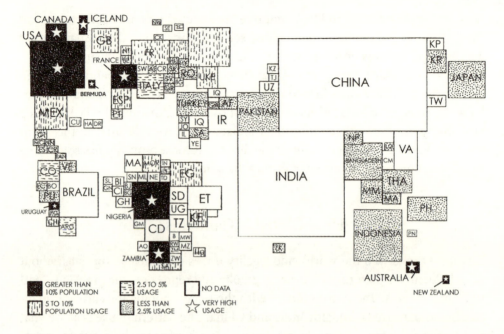

Figure 17.7. Global distribution of marijuana use. Dark rectangles indicate countries with greater than 10 percent usage among the population; vertical stripes indicate between 5 and 10 percent population use; horizontal stripes indicate 2.5 to 5 percent use; and stipples indicate less than 2.5 percent use. White indicates no data available for that country. The stars indicate nine countries with very high marijuana use: Iceland (18.3 percent), US (16.2 percent), Nigeria (14.3 percent), Canada (12.7 percent), France (11.1 percent), New Zealand (11 percent), Australia (10.2 percent), Zambia (9.5 percent), and Uruguay (9.3 percent). Drawing by Rob DeSalle and Patricia Wynne.

survey was taken, and this most assuredly influenced how marijuana is used in that country. Fines for marijuana use in Iceland are somewhat strict, and it is still not legal at the time of this writing.

This leaves the United States as the world's number one pot-smoking country. According to a study by the Cannabis Price Index, New York City, Los Angeles, and Chicago are three of the top ten cities with the highest consumption of cannabis. New York City consumes almost twice of what the next closest city, Karachi, does per year. The other cities in the top ten are London, Moscow, Mumbai, New Delhi, Cairo, and Toronto. One needs to be careful in interpreting the results of this comparison, though, because the top ten cannabis-using cities are also among the top largest cities in population size.

Different societies take different views of marijuana's place in their culture. Icelanders took some offense at being identified by the study in figure 17.7 as the number one pot-smoking country. Other nations celebrate their pot-smoking heritage, and there is no more celebratory medium for cannabis than film, which we turn to for our next chart.

Movies and Pot

Since the early twentieth century, cinema has become an important part of many cultures. In 1999 a writer named Kip Kay published an annotated list of all the movies between 1930 and 1999 that showed or referred to marijuana use. This must have been an entertaining exercise, because he covers over sixty years and 550 films in his list. Figure 17.8 plots the number of movies each year that included or referred to marijuana use.

Few films referenced marijuana before the 1960s. The ones that did were often cautionary tales against getting high. Because of its poor acting, sensationalism, and outright bizarreness, *Reefer Madness* is considered the grandfather of all marijuana movies. Released in 1936 as part of Anslinger's Army's arsenal to educate the public about the evils of marijuana, it was not, however, the first movie to reference marijuana. Two years before the release of *Reefer Madness, Murder at the Vanities* was released. This film couldn't decide whether it was a comedy, musical, detective story, mobster movie, or drama, but it did feature B-movie actress Gertrude Michael singing "Sweet Marijuana" with the Duke Ellington Band. The song's refrain ("You alone can bring my lover back to me / Even though I know it's all a fantasy / And then put me to sleep / Sweet Marijuana, Marijuana") suggested a dreamy but sad relationship between cannabis and the singer.

It is easy to imagine that the long gap between *Reefer Madness* and several 1950s movies referencing marijuana was the result of the craziness about cannabis engendered by that movie. Perhaps *Reefer Madness* was effective propaganda (the "young" people in the movie either die, go crazy, or are arrested for murder), or perhaps marijuana use wasn't an issue during those years. People in the United States had bigger concerns, including recovery from a crippling economic depression and the Second World War. There are two movies that are especially interesting from the 1950s: *High School Confidential* and *The Gene Krupa Story*. The latter film tells the story of showy jazz drummer Krupa's marijuana conviction (and drumstick dropping due to marijuana use). As the movie dramatically reveals,

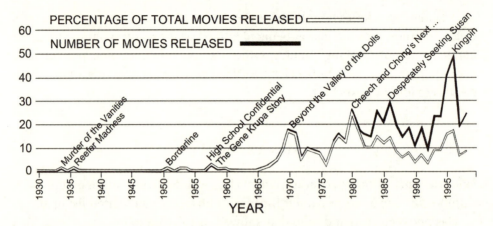

Figure 17.8. Percentage of movies referencing marijuana released by year (white) and number of movies released by year (black). Films from the early and peak years are listed.

Krupa served time but redeemed himself. The movie is almost a paean to drug use by musicians. *High School Confidential,* while slightly better acted than *Reefer Madness,* takes a similar stand on marijuana use by teens. I use the word "teens" lightly, because the actors in both *Reefer Madness* and *High School Confidential* appear to be in their late twenties to early thirties. The message of both movies was that bad things happen when you smoke marijuana.

Then came the late sixties and an infusion of movies using marijuana as a plot device. There was a large increase in films with marijuana-tinged themes throughout the seventies, eighties, and nineties. Between 1970 and 1998 an average of about twenty movies per year used marijuana in their storylines; the year with the highest number of marijuana films was 1996, with forty-seven movies. Although this is a high result, if the number is adjusted relative to the total number of films released in a year, 1980 appears to be the most saturated year before 1999, which is the extent of Kay's survey. Many of these movies incorporated marijuana in a light, humorous way that mirrored the relaxed attitudes toward marijuana of those decades.

The twenty-first century ushered in a freedom to incorporate marijuana into storylines in ways not previously imagined. It is, however, difficult to estimate the number of films referring to marijuana since the 1990s, because movie releases multiplied many times from then to the present. On the other hand, a search of the internet reveals many "top ten" lists of movies referencing marijuana that be-

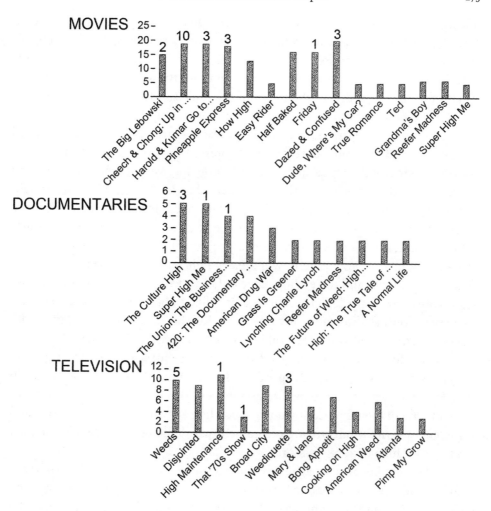

Figure 17.9. Compiled internet rankings of movies, documentaries, and television series that reference marijuana.

came popular in the early twenty-first century. Figure 17.9 summarizes thirty top ten lists. I have ranked the movies in these lists by their position in the lists and normalized their rank by the number of lists they appear in. The numbers above the bars in the figure refer to the number of lists in which a particular film was ranked number one. The movies in the figure are ranked from best (on the left) to next best (to the right). A total of 141 movies appear on the thirty top ten lists, and

The Big Lebowski was ranked at the top using the criteria established for overall ranking. Although it appears as number one in only two lists, it is found on almost everyone's top ten list. The movie with the most number one rankings is Cheech and Chong's *Up in Smoke,* with ten. It is not the ultimate number one because of the strong reaction about half of the rankers had against it. Either they loved it and ranked it number one, or hated it and left it off their lists.

I consider these two films among the best of marijuana movies myself, and *The Big Lebowski* certainly deserves the highest of rankings. This Coen brothers movie is a kind of detective story, a nod to bowling, very much a comedy, and definitely an ode to marijuana. The Dude (played by Jeff Bridges) abides throughout the movie, and there are two trippy dream sequences that nicely detail the marijuana high. One of the sequences, induced by a sock to the Dude's jaw, starts off with colorful exploding fireworks and progresses to the Dude flying through space Superman style, with a bowling ball in his outstretched right hand. The fluidity of his flight and the ending panic of falling from the sky and abrupt waking up are convincing depictions of what it is like to be high. The second dream sequence is only slightly more down to earth. Bowling again is a part of the experience, and the overall dream resembles a 1930s Busby Berkeley movie. The image of Saddam Hussein handing the Dude his bowling shoes and the First Edition's hazy background song, "Just Dropped In (To See What Condition My Condition Was In)," are perfect marijuana dream devices. This dream sequence contrasts the euphoria of the high (sliding down a bowling alley between the legs of dancing women) with the unceremonious comedown, depicted by three red-clad men chasing the Dude with big scissors.

Television has also broached marijuana use in many ways. The rise of cable television and pay networks such as Home Box Office and Showtime pushed the envelope, with the Showtime series *Weeds* ranking as number one over all of the ten lists summarized in figure 17.9. This series about a Southern California widow building a marijuana sales empire was rated number one on half the top ten lists and was universally regarded as a top ten series in the rankings. One network series that ranked fairly high was *That '70s Show,* with its now famous circle smoking scenes. Censorship on network television has been stricter than for the cable and streaming channels that prevail today, and these scenes were somewhat controversial when they first appeared on the Fox network. Prohibited from showing marijuana smoking on-screen because of censors, the producers decided to set a camera at the center of a table that would rotate to view the individuals sitting at the table (but without a joint), who would then make stoned remarks to the camera. As the joint gets passed around the table, the actual act of smoking was avoided. *That '70s Show* (airing in the late 1990s) has given rise to *That '90s Show* (airing in 2023–2024), where the children of the teens in the seventies show are now the subjects. The rotating camera device proves as effective today in the new series produced by Netflix as it was in the late 1990s for the original series.

The TV top ten list includes cooking shows such as *Bong Appetit* and *Cooking on High.* Julia Child was perhaps the first to produce cooking shows for television, her *French Chef* series airing initially in the 1960s. Since her pioneering work, cooking shows have become part of the culture of many countries. A nation's cuisine is integral to its culture, and cooking has always been a part of the culture and draw of marijuana. It appears that marijuana has made inroads into this mainstream way of entertaining and educating people about food.

My favorite on the top ten lists for marijuana documentaries is *Super High Me,* made in 2007. It follows comedian Doug Benson, who abstains from marijuana for thirty days and then abruptly stays stoned on marijuana for another thirty days. During both periods he is examined by a physician, given tests for his sperm count, and takes comparative scholastic tests. This movie could be considered a send-up of the 2004 documentary called *Super Size Me,* in which filmmaker Morgan Spurlock subsisted on fast food for thirty days. His physical condition was assessed at the end by a physician who determined that Spurlock had gained weight and by all other measures of health had deteriorated. But in *Super High Me* cannabis use as assessed by the physician in the film has had minimal if not negligible impact on Benson's health. Ironically, Morgan Spurlock passed away in 2024 at the age of fifty-four. His monthlong binge on fast food may have had something to do with that. We will have to wait and see whether Benson's long-term health reveals anything about cannabis usage and its contribution to lifespan.

Epilogue

The Message

C annabis is a fascinating plant and has a significant place in our planet's magnificent tree of life, replete with unusual ancestors and remarkable ecological interactions with other plants and animals. It is a complex organism with many hundreds of chemical compounds coursing through its seeds, stems, leaves, and flowers. Its impact on our brains when we use it is equally stupefying and intricate. And it has a complex sex life that we humans have learned to manipulate to make it easier for us to use it in industrial, cultural, medicinal, and recreational ways. Marijuana also has a complex genetic makeup, and the ways that we use it require that we understand its genetic architectures. Cannabis has profoundly influenced our species culturally and socially, and continues to do so as we grapple with the complex legal aspects of its use.

Cannabis has inserted itself into our modern culture in many ways. It also appears that attitudes to cannabis use seem to be relaxing, despite remaining illegal in four states in the United States and a number of countries throughout the world. Marijuana has also become a major consideration for legal and social science thinkers. It has become a major consideration in the economics of US states and other nations. Although we are seeing a relaxation of adverse attitudes to marijuana, cannabis is still viewed negatively by various policymakers and governments. This negative attitude is best exemplified by the focus of research money away from medical considerations and toward funding investigations of its harmfulness. Understanding the harmful aspects of any drug is of course critical to demonstrating the utility of that drug as medicine; assessing the dangers of a medical treatment is a part of the clinical trial process. But it remains surprising that the benefits and uses of cannabis don't demand more of funders' attention.

To lend some perspective, consider Lester Grinspoon and James Bakalar's plea for more research on cannabis and a more rational approach when considering marijuana as a part of our everyday lives. They made this plea about thirty years ago, and it is only in the past few years that attitudes in the US have relaxed. Our tolerance for marijuana use is increasing, and we are finding better ways to resolve some of the harmful aspects.

Lester Grinspoon, who passed away at the age of ninety-two in 2020, was not a lifetime advocate for marijuana use. His initial attitude was that it might be harmful. But after being convinced by Carl Sagan to try it, he became a determined advocate for using marijuana in medicine. He and Bakalar documented many of the ways cannabis could serve as a medicinal. He did most of this campaigning during the cannabis "dark ages" in the US, when marijuana was considered a dangerous narcotic and many people were incarcerated for possessing even small amounts of it. It is through Grinspoon's diligence that much of the medical context of cannabis was clarified. He also had a lot to do with the changing social attitudes.

One of the dreams of any biologist is to have a species or natural law named after them. I soberingly point out that I have an insignificant little fly named after me (*Zygothrica desallei*), whereas Charles Darwin had over 250 species of charismatic plants and beautiful animals named after him. But what could be better than to have a strain of marijuana named after you?

Charlotte's Web is a strain of cannabis with high CBD content and low THC content (characterized at 0.5 percent THC and 17 percent CBD). Originally called Hippie's Disappointment (due to the low THC content and lack of sufficient kick), it was renamed in honor of a Colorado Springs girl, Charlotte Figi, in the 2010s. Charlotte suffered from grand mal epileptic seizures throughout her infancy. Her parents searched for a cannabis strain with the very characteristics of Hippie's Disappointment for treating the debilitating epileptic effects of the syndrome called severe myoclonic epilepsy of infancy or SMEI. She suffered up to

300 grand mall seizures per week, and the oil from the strain helped her significantly. She died in 2020 at the age of fourteen but is remembered as the "girl who changed marijuana laws." The Hippie's Disappointment strain was renamed in her honor to Charlotte's Web, with reference to her first name and to the popular children's story written by E. B. White and illustrated by Garth Williams, which first appeared in 1952. This lovely story about the relationship of a barn spider named Charlotte and a young pig named Wilbur has been beloved by several generations of children. Renaming the strain of marijuana that had helped Charlotte Figi was an homage to her and her role in the marijuana story.

Lester Grinspoon has a strain of marijuana named after him called Dr. Grinspoon. One marijuana information website describes the strain as follows: "[Dr. Grinspoon] is found to be a good strain for patients who are fighting mental disorders and will be a fine mood elevator. Moreover, Dr. Grinspoon would help in alleviating mood disorders like attention deficit hyperactivity disorder and pain." Hence it was perfectly appropriate to name this strain after him. I strongly believe that if we take a Grinspoonian approach to cannabis and its incorporation into our medical toolbox, we will be much better off in the future. This approach will also connect the past cultural context of cannabis with its present one, and better inform us of the role of cannabis in our society.

For Further Reading

CHAPTER 1. Drunken Monkeys and Stoned Spiders

For a discussion of the drunken monkey hypothesis, see Robert Dudley's 2014 book *The Drunken Monkey: Why We Drink and Abuse Alcohol*, Ian Tattersall's and my book on spirits, and Hockings and Dunbar's book on alcohol consumption in humans. The distribution of alcohol tolerance and use in primates is described in Amato et al. (2021). The use of spider webs as an indicator of neural togetherness is discussed in Noever et al. (1995). Insect coexistence with cannabis is discussed in several papers listed below (De Petrocellis et al. 1999; Batra 1976; Crosby et al. 1986; and the three papers by John McPartland). Catnip intoxication is described in Petersik et al. (1969) and Janeczek et al. (2018). Peter Witt's classic 1954 paper on stoned spiders is also mentioned here.

Amato, Katherine R., Óscar M. Chaves, Elizabeth K. Mallott, Timothy M. Eppley, Filipa Abreu, Andrea L. Baden, Adrian A. Barnett, et al. "Fermented Food Consumption in Wild Nonhuman Primates and Its Ecological Drivers." *American Journal of Physical Anthropology* 175, no. 3 (2021): 513–530.

Batra, S. W. T. "Some Insects Associated with Hemp or Marijuana (*Cannabis sativa* L.) in Northern India." *Journal of the Kansas Entomological Society* 49, no. 3 (1976): 385–388.

Crosby, T. K., J. C. Watt, A. C. Kistemaker, and P. E. Nelson. "Entomological Identification of the Origin of Imported Cannabis." *Journal of the Forensic Science Society* 26, no. 1 (1986): 35–44.

De Petrocellis, L., D. Melck, T. Bisogno, A. Milone, and V. Di Marzo. "Finding of the Endocannabinoid Signalling System in Hydra, a Very Primitive Organism: Possible Role in the Feeding Response." *Neuroscience* 92, no. 1 (1999): 377–387.

DeSalle, Rob, and Ian Tattersall. *Distilled: A Natural History of Spirits.* Yale University Press, 2022.

Dudley, Robert. *The Drunken Monkey: Why We Drink and Abuse Alcohol.* University of California Press, 2014.

Hockings, Kimberley, and Robin Dunbar, eds. *Alcohol and Humans: A Long and Social Affair.* Oxford University Press, 2019.

Jackson, Basil, and Alan Reed. "Catnip and the Alteration of Consciousness." *Journal of the American Medical Association* 207, no. 7 (1969): 1349–1350.

Janeczek, Agnieszka, Marcin Zawadzki, Pawel Szpot, and Artur Niedzwiedz. "Marijuana Intoxication in a Cat." *Acta Veterinaria Scandinavica* 60, no. 1 (2018): 1–4.

McPartland, John M. "Cannabis Pests." *Journal of the International Hemp Association* 3, no. 2 (1996): 52–55.

McPartland, John, Vincenzo Di Marzo, Luciano De Petrocellis, Alison Mercer, and Michelle Glass. "Cannabinoid Receptors Are Absent In Insects." *Journal of Comparative Neurology* 436, no. 4 (2001): 423–429.

McPartland, J. M., J. Agraval, D. Gleeson, K. Heasman, and M. Glass. "Cannabinoid Receptors in Invertebrates." *Journal of Evolutionary Biology* 19, no. 2 (2006): 366–373.

Noever, David A., Raymond J. Cronise, and Rachna A. Relwani. "Using Spider-Web Patterns to Determine Toxicity." *NASA Tech Briefs* 19, no. 4 (1995): 82.

Petersik, John T., John Poundstone, J. Worth Estes, Sidney Cohen, Dudley M. King, Philip G. Seitner, John H. Tanton, et al. "Of Cats, Catnip, and Cannabis." *Journal of the American Medical Association* 208, no. 2 (1969): 360.

Witt, Peter. "Spider Webs and Drugs." *Scientific American* 191, no. 6 (1954): 80–87.

CHAPTER 2. Cannabis: A Short Cultural History

The Clarke and Merlin tome is the basis for any cogent discussion of cannabis and culture.

Clarke, Robert C., and Mark D. Merlin. *Cannabis: Evolution and Ethnobotany.* University of California Press, 2013.

Vera Rubin's *Cannabis and Culture* volume reference is also pertinent.

Rubin, Vera, ed. *Cannabis and Culture.* Mouton Publishers, 1975.

The following literature documents much of the cultural context of cannabis throughout modern human history, starting about 12,000 to 10,000 years ago.

Arie, Eran, Baruch Rosen, and Dvory Namdar. "Cannabis and Frankincense at the Judahite Shrine of Arad." *Tel Aviv* 47, no. 1 (2020): 5–28.

Pisanti, Simona, and Maurizio Bifulco. "Medical Cannabis: A Plurimillennial History of an Evergreen." *Journal of Cellular Physiology* 234, no. 6 (2019): 8342–8351.

Russo, Ethan B. "History of Cannabis and Its Preparations in Saga, Science, and Sobriquet." *Chemistry & Biodiversity* 4, no. 8 (2007): 1614–1648.

These papers can be consulted for perspectives on the archaeological evidence for cannabis associations with humans in Asia.

Kobayashi, M., A. Momohara, S. Okitsu, S. Yanagisawa, and T. Okamoto. "Fossil Hemp Fruits in the Earliest Jomon Period from the Okinoshima Site, Chiba Prefecture." *Shokuseishi kenkyū* 16, no. 1 (2008): 11–18.

McPartland, John M., William Hegman, and Tengwen Long. "Cannabis in Asia: Its Center of Origin and Early Cultivation, Based on a Synthesis of Subfossil Pollen and Archaeobotanical Studies." *Vegetation History and Archaeobotany* 28, no. 6 (2019): 691–702.

Okazaki, Hiroko, Makiko Kobayashi, Arata Momohara, Sei-ichi Eguchi, Tozo Okamoto, Sei-ichi Yanagisawa, Susumu Okubo, and Jota Kiyonaga. "Early Holocene Coastal Environment Change Inferred from Deposits at Okinoshima Archeological Site, Boso Peninsula, Central Japan." *Quaternary International* 230, no. 1–2 (2011): 87–94.

Bronze-age cannabis association in Eurasia and western Europe are documented in the following papers.

Fleming, Michael P., and Robert C. Clarke. "Physical Evidence for the Antiquity of *Cannabis sativa* L." *Journal of the International Hemp Association* 5, no. 2 (1998): 80–95.

Long, Tengwen, Mayke Wagner, Dieter Demske, Christian Leipe, and Pavel E. Tarasov. "Cannabis in Eurasia: Origin of Human Use and Bronze Age Trans-Continental Connections." *Vegetation History and Archaeobotany* 26, no. 2 (2017): 245–258.

McPartland, John M., Geoffrey W. Guy, and William Hegman. "Cannabis Is Indigenous to Europe and Cultivation Began during the Copper or Bronze Age: A Probabilistic Synthesis of Fossil Pollen Studies." *Vegetation History and Archaeobotany* 27, no. 4 (2018): 635–648.

Ren, Meng, Zihua Tang, Xinhua Wu, Robert Spengler, Hongen Jiang, Yimin Yang, and Nicole Boivin. "The Origins of Cannabis Smoking: Chemical Residue Evidence from the First Millennium BCE in the Pamirs." *Science Advances* 5, no. 6 (2019): eaaw1391.

Sumler, Alan. *Cannabis in the Ancient Greek and Roman World.* Lexington Books, 2018.

Worldwide perspectives are discussed in the following references.

Chouvy, Pierre-Arnaud. "Cannabis Cultivation in the World: Heritages, Trends and Challenges." *EchoGéo* 48 (2019).

Warf, Barney. "High Points: An Historical Geography of Cannabis." *Geographical Review* 104, no. 4 (2014): 414–438.

The uses of cannabis as fiber are detailed in these sources.

Balant, Manica, Airy Gras, Francisco Gálvez, Teresa Garnatje, Joan Vallès, and Daniel Vitales. Daniel. CANNUSE, the Database of Traditional Uses. https://doi.org /10.20350/digitalCSIC/13686 (2021).

Fike, John. "The History of Hemp." In *Industrial Hemp as a Modern Commodity Crop*, edited by D. W. Williams, 1–25. ASA, CSSA, and SSSA Books, 2019.

Lavrieux, Marlène, Jérémy Jacob, Jean-Robert Disnar, Jean-Gabriel Bréhéret, Claude Le Milbeau, Yannick Miras, and Valérie Andrieu-Ponel. "Sedimentary Cannabinol Tracks the History of Hemp Retting." *Geology* 41, no. 7 (2013): 751–754.

Cannabis as an early psychoactive is discussed in detail in the following.

Jiang, Hongen, Long Wang, Mark D. Merlin, Robert C. Clarke, Yan Pan, Yong Zhang, Guoqiang Xiao, and Xiaolian Ding. "Ancient Cannabis Burial Shroud in a Central Eurasian Cemetery." *Economic Botany* 70, no. 3 (2016): 213–221.

Nahas, Gabriel G. "Hashish in Islam 9th to 18th Century." *Bulletin of the New York Academy of Medicine* 58, no. 9 (1982): 814–831.

Ren, Meng, Zihua Tang, Xinhua Wu, Robert Spengler, Hongen Jiang, Yimin Yang, and Nicole Boivin. "The Origins of Cannabis Smoking: Chemical Residue Evidence from the First Millennium BCE in the Pamirs." *Science Advances* 5, no. 6 (2019): eaaw1391.

CHAPTER 3. Origins

The genetic advantage of having sex over being asexual was discussed in the 1960s in the following paper.

Crow, James F., and Motoo Kimura. "Evolution in Sexual and Asexual Populations." *American Naturalist* 99, no. 909 (1965): 439–450.

I have referenced several of John McPartland's papers throughout this chapter and elsewhere. He is the reigning expert on cannabis, and his work includes the following publications.

McPartland, J. "Cannabis: The Plant, Its Evolution, and Its Genetics—with an Emphasis on Italy." *Rendiconti Lincei: Scienze Fisiche e Naturali* 31, no. 4 (2020): 939–948.

McPartland, John M. "Cannabis Systematics at the Levels of Family, Genus, and Species." *Cannabis and Cannabinoid Research* 3, no. 1 (2018): 203–212.

McPartland, John M., and Geoffrey W. Guy. "Models of Cannabis Taxonomy, Cultural Bias, and Conflicts between Scientific and Vernacular Names." *Botanical Review* 83, no. 4 (2017): 327–381.

McPartland, John M., and Ernest Small. "A Classification of Endangered High-THC Cannabis (*Cannabis sativa* subsp. *indica*) Domesticates and Their Wild Relatives." *PhytoKeys* 144 (2020): 81–112.

The early oxygen status of the earth is discussed at length in the following two publications.

Biello, David. "The Origin of Oxygen in Earth's Atmosphere." *Scientific American,* August 19, 2009.

Lane, Nick. *Oxygen: The Molecule That Made the World.* Oxford University Press, 2002.

The best reference for the evolution of sex is still John Maynard Smith's classic work.

Smith, John Maynard. *The Evolution of Sex.* Cambridge University Press, 1978.

Animal relationships and phylogenetic patterns are discussed at length in the following source.

Neumann, Johannes S., Michael Tessler, Rob DeSalle, and Bernd Schierwater. "Modern Invertebrate Systematics: The Phylogenetics of Early Metazoa." In *Invertebrate Zoology: A Tree of Life Approach,* edited by Bernd Schierwater and Rob DeSalle, 65–80. CRC Press, 2021.

Algae and chloroplast genomes are detailed in the following paper.

Maruyama, Shinichiro, and Eunsoo Kim. "A Modern Descendant of Early Green Algal Phagotrophs." *Current Biology* 23, no. 12 (2013): 1081–1084.

References on plant evolution, plant fossils, and the evolution of flowers can be found in the following literature.

Benton, Michael J., Peter Wilf, and Hervé Sauquet. "The Angiosperm Terrestrial Revolution and the Origins of Modern Biodiversity." *New Phytologist* 233, no. 5 (2022): 2017–2035.

Coiro, Mario, James A. Doyle, and Jason Hilton. "How Deep Is the Conflict between Molecular and Fossil Evidence on the Age of Angiosperms?" *New Phytologist* 223, no. 1 (2019): 83–99.

Fu, Qiang, Jose Bienvenido Diez, Mike Pole, Manuel García Ávila, Zhong-Jian Liu, Hang Chu, Yemao Hou, et al. "An Unexpected Noncarpellate Epigynous Flower from the Jurassic of China." *Elife* 7 (2018): e38827.

Li, Hong-Tao, Ting-Shuang Yi, Lian-Ming Gao, Peng-Fei Ma, Ting Zhang, Jun-Bo Yang, Matthew A. Gitzendanner, et al. "Origin of Angiosperms and the Puzzle of the Jurassic Gap." *Nature Plants* 5, no. 5 (2019): 461–470.

Sauquet, Hervé, Santiago Ramírez-Barahona, and Susana Magallón. "The Age of Flowering Plants Is Unknown." *EcoEvoRxiv,* https://doi.org/10.32942/osf.io/n4v6b (2022).

Sauquet, Hervé, Maria Von Balthazar, Susana Magallón, James A. Doyle, Peter K. Endress, Emily J. Bailes, Erica Barroso de Morais, et al. "The Ancestral Flower of Angiosperms and Its Early Diversification." *Nature Communications* 8, no. 1 (2017): 1–10.

Silvestro, Daniele, Christine D. Bacon, Wenna Ding, Qiuyue Zhang, Philip C. J. Donoghue, Alexandre Antonelli, and Yaowu Xing. "Fossil Data Support a Pre-Cretaceous Origin of Flowering Plants." *Nature Ecology & Evolution* 5, no. 4 (2021): 449–457.

More information on the cannabis taxonomic problem can be found in the following.

Pollio, Antonino. "The Name of Cannabis: A Short Guide for Nonbotanists." *Cannabis and Cannabinoid Research* 1, no. 1 (2016): 234–238.

Schräder, Nicholas H. B., Marieke C. Bolling, and André P. Wolff. "Letter to the Editor: Ordering the Chaos in Cannabinoid-Related Research: Is It Time for a Task Force on Taxonomy?" *Cannabis and Cannabinoid Research* 6, no. 2 (2021): 174–175.

CHAPTER 4. Hops and Hemp

John McPartland has greatly influenced the systematic discussion of cannabis. References for his work in this chapter are as follows.

McPartland, J. "Cannabis: The Plant, Its Evolution, and Its Genetics—with an Emphasis on Italy." *Rendiconti Lincei: Scienze Fisiche e Naturali* 31, no. 4 (2020): 939–948.

McPartland, John M. "Cannabis Systematics at the Levels of Family, Genus, and Species." *Cannabis and Cannabinoid Research* 3, no. 1 (2018): 203–212.

McPartland, John M., and Geoffrey W. Guy. "Models of Cannabis Taxonomy, Cultural Bias, and Conflicts between Scientific and Vernacular Names." *Botanical Review* 83, no. 4 (2017): 327–381.

McPartland, John M., and Ernest Small. "A Classification of Endangered High-THC Cannabis (*Cannabis sativa* subsp. *indica*) Domesticates and Their Wild Relatives." *PhytoKeys* 144 (2020): 81–112.

The following are references on the spread of cannabis.

Charitos, Ioannis A., Roberto Gagliano-Candela, Luigi Santacroce, and Lucrezia Bottalico. "The Cannabis Spread throughout the Continents and Its Therapeutic Use in History." *Endocrine, Metabolic & Immune Disorders—Drug Targets* 21, no. 3 (2021): 407–417.

Yang, Mei-Qing, Robin van Velzen, Freek T. Bakker, Ali Sattarian, De-Zhu Li, and Ting-Shuang Yi. "Molecular Phylogenetics and Character Evolution of Cannabaceae." *Taxon* 62, no. 3 (2013): 473–485.

Zuardi, Antonio Waldo. "History of Cannabis as a Medicine: A Review." *Brazilian Journal of Psychiatry* 28, no. 2 (2006): 153–157.

Taxonomic problems in cannabis are discussed in the following sources.

Pollio, Antonino. "The Name of Cannabis: A Short Guide for Nonbotanists." *Cannabis and Cannabinoid Research* 1, no. 1 (2016): 234–238.

Schräder, Nicholas H. B., Marieke C. Bolling, and André P. Wolff. "Letter to the Editor: Ordering the Chaos in Cannabinoid-Related Research: Is It Time for a Task Force on Taxonomy?" *Cannabis and Cannabinoid Research* 6, no. 2 (2021): 174–175.

Schultes, Richard Evans, William M. Klein, Timothy Plowman, and Tom E. Lockwood. "Cannabis: an example of taxonomic neglect." In *Cannabis and Culture,* edited by Vera Rubin, 21–38. Mouton Publishers, 1975.

CHAPTER 5. A Complicated Sex Life

The biochemistry and vegetative parts of cannabis are discussed in detail in the following biochemistry publications.

Aiello, Gilda, Elisa Fasoli, Giovanna Boschin, Carmen Lammi, Chiara Zanoni, Attilio Citterio, and Anna Arnoldi. "Proteomic Characterization of Hempseed (*Cannabis sativa* L.)." *Journal of Proteomics* 147 (2016): 187–196.

Cattaneo, Chiara, Annalisa Givonetti, Valeria Leoni, Nicoletta Guerrieri, Marcello Manfredi, Annamaria Giorgi, and Maria Cavaletto. "Biochemical Aspects of Seeds from *Cannabis sativa* L. Plants Grown in a Mountain Environment." *Scientific Reports* 11, no. 1 (2021): 1–19.

Kakabouki, Ioanna, Alexandros Tataridas, Antonios Mavroeidis, Angeliki Kousta, Stella Karydogianni, Charikleia Zisi, Varvara Kouneli, et al. "Effect of Colonization of *Trichoderma harzianum* on Growth Development and CBD Content of Hemp (*Cannabis sativa* L.)." *Microorganisms* 9, no. 3 (2021): 518.

Mamone, Gianfranco, Gianluca Picariello, Alessia Ramondo, Maria Adalgisa Nicolai, and Pasquale Ferranti. "Production, Digestibility and Allergenicity of Hemp (*Cannabis sativa* L.) Protein Isolates." *Food Research International* 115 (2019): 562–571.

Park, Seul-Ki, Jong-Bok Seo, and Mi-Young Lee. "Proteomic Profiling of Hempseed Proteins from Cheungsam." *Biochimica et Biophysica Acta (BBA)—Proteins and Proteomics* 1824, no. 2 (2012): 374–382.

Plant sex, plant genitals, and plant mating are covered in the following articles, some of which concern plants in general and some of which focus on cannabis. I also include a short reminder that plants do not have brains like ours.

DeSalle, Rob, and Ian Tattersall. "Do Plants Have Brains?" *Natural History,* November 2008.

Islam, Mohammad Moinul, Zed Rengel, Paul Storer, Kadambot H. M. Siddique, and Zakaria M. Solaiman. "Industrial Hemp (*Cannabis sativa* L.) Varieties and Seed Pre-Treatments Affect Seed Germination and Early Growth of Seedlings." *Agronomy* 12, no. 1 (2021): 6.

Karsten, H. "IX.—On the Sexual Life of Plants, and Parthenogenesis." *Annals and Magazine of Natural History* 8, no. 44 (1861): 81–99.

Parsons, Jessica L., Sara L. Martin, Tracey James, Gregory Golenia, Ekaterina A. Boudko, and Shelley R. Hepworth. "Polyploidization for the Genetic Improvement of *Cannabis sativa.*" *Frontiers in Plant Science* 10 (2019): 476.

Punja, Zamir K., and Janesse E. Holmes. "Hermaphroditism in Marijuana (*Cannabis sativa* L.) Inflorescences—Impact on Floral Morphology, Seed Formation, Progeny Sex Ratios, and Genetic Variation." *Frontiers in Plant Science* 11 (2020): 718.

Renner, Susanne S. "The Relative and Absolute Frequencies of Angiosperm Sexual Systems: Dioecy, Monoecy, Gynodioecy, and an Updated Online Database." *American Journal of Botany* 101, no. 10 (2014): 1588–1596.

Vijverberg, Kitty, Peggy Ozias-Akins, and M. Eric Schranz. "Identifying and Engineering Genes for Parthenogenesis in Plants." *Frontiers in Plant Science* 10 (2019): 128.

CHAPTER 6. Buds

I highly recommend the following two of Darwin's classic barnacle papers. They are amazing in their scope, and the writing is beautiful.

Darwin, C. R. *Monograph on the Sub-Class Cirripedia: The Lepadidae; or, Pedunculated Cirripedes* (1851), F339.1.

Darwin, C. R. *Monograph on the Sub-Class Cirripedia: The Balanidae (or Sessile Cirripedes); The Verrucidae, etc.* (1854), F339.2.

Plants and the fossil record are covered in the following articles from the scientific literature. Some are dense, but most are good fun to read.

Bateman, Richard M. "Hunting the Snark: The Flawed Search for Mythical Jurassic Angiosperms." *Journal of Experimental Botany* 71, no. 1 (2020): 22–35.

Cui, Da-Fang, Yemao Hou, Pengfei Yin, and Xin Wang. "A Jurassic Flower Bud from China." *Geological Society, London, Special Publications* 521 (2022).

Harris, Liam W., and T. Jonathan Davies. "A Complete Fossil-Calibrated Phylogeny of Seed Plant Families as a Tool for Comparative Analyses: Testing the 'Time For Speciation' Hypothesis." *PloS One* 11, no. 10 (2016): e0162907.

Janssens, Steven B., Thomas L. P. Couvreur, Arne Mertens, Gilles Dauby, Leo-Paul M. J. Dagallier, Samuel Vanden Abeele, Filip Vandelook, et al. "A Large-Scale Species Level Dated Angiosperm Phylogeny for Evolutionary and Ecological Analyses." *Biodiversity Data Journal* 8 (2020): e39677.

Morris, Jennifer L., Mark N. Puttick, James W. Clark, Dianne Edwards, Paul Kenrick, Silvia Pressel, Charles H. Wellman, Ziheng Yang, Harald Schneider, and Philip C. J. Donoghue. "The Timescale of Early Land Plant Evolution." *Proceedings of the National Academy of Sciences* 115, no. 10 (2018): E2274–E2283.

Silvestro, Daniele, Christine D. Bacon, Wenna Ding, Qiuyue Zhang, Philip C. J. Donoghue, Alexandre Antonelli, and Yaowu Xing. "Fossil Data Support a Pre-Cretaceous Origin of Flowering Plants." *Nature Ecology & Evolution* 5, no. 4 (2021): 449–457.

Wang, Xin. "A Novel Early Cretaceous Flower and Its Implications on Flower Derivation." *Biology* 11, no. 7 (2022): 1036.

The following two citations address floral development from the historical perspective.

Dornelas, Marcelo Carnier, and Odair Dornelas. "From Leaf to Flower: Revisiting Goethe's Concepts on the 'Metamorphosis' of Plants." *Brazilian Journal of Plant Physiology* 17, no. 4 (2005): 335–344.

Richards, Robert J. "The Foundations of Archetype Theory in Evolutionary Biology: Kant, Goethe, and Carus." *Republics of Letters* 6, no. 1 (2018).

These publications address the ABC (ABCE) model of floral development.

Alvarez-Buylla, Elena R., Mariana Benítez, Adriana Corvera-Poiré, Álvaro Chaos Cador, Stefan de Folter, Alicia Gamboa de Buen, Adriana Garay-Arroyo, et al. "Flower Development." *Arabidopsis Book* 8 (2010): e0127.

Bowman, John L., David R. Smyth, and Elliot M. Meyerowitz. "The ABC Model of Flower Development: Then and Now." *Development* 139, no. 22 (2012): 4095–4098.

Chanderbali, Andre S., Brent A. Berger, Dianella G. Howarth, Pamela S. Soltis, and
    Douglas E. Soltis. "Evolving Ideas on the Origin and Evolution of Flowers: New
    Perspectives in the Genomic Era." *Genetics* 202, no. 4 (2016): 1255–1265.
Irish, Vivian. "The ABC Model of Floral Development." *Current Biology* 27, no. 17 (2017):
    R887–R890.
Litt, Amy, and Elena M. Kramer. "The ABC Model and the Diversification of Floral
    Organ Identity." *Seminars in Cell & Developmental Biology* 21, no. 1 (2010): 129–137.
Zahn, Laura M., Baomin Feng, and Hong Ma. "Beyond the ABC-Model: Regulation of
    Floral Homeotic Genes." *Advances in Botanical Research* 44 (2006): 163–207.

The following articles concern cannabis flower development specifically.

Leme, Flávia M., Jürg Schönenberger, Yannick M. Staedler, and Simone P. Teixeira.
    "Comparative Floral Development Reveals Novel Aspects of Structure and Diver-
    sity of Flowers in Cannabaceae." *Botanical Journal of the Linnean Society* 193, no. 1
    (2020): 64–83.
Leme, Flávia M., Yannick M. Staedler, Jürg Schönenberger, and Simone P. Teixeira.
    "Floral Morphogenesis of *Celtis* Species: Implications for Breeding System and
    Reduced Floral Structure." *American Journal of Botany* 108, no. 9 (2021): 1595–1611.
Punja, Zamir K., and Janesse E. Holmes. "Hermaphroditism in Marijuana (*Cannabis
    sativa* L.) Inflorescences—Impact on Floral Morphology, Seed Formation, Prog-
    eny Sex Ratios, and Genetic Variation." *Frontiers in Plant Science* 11 (2020): 718.
Spitzer-Rimon, Ben, Shai Duchin, Nirit Bernstein, and Rina Kamenetsky. "Architecture
    and Florogenesis in Female *Cannabis sativa* Plants." *Frontiers in Plant Science* 10
    (2019): 350.

The quote from Goethe appears in a letter of 1787 to J. G. Herder, quoted in the following
volume.

Goethe, Johann Wolfgang von. *Goethe's Botanical Writings*. Translated by Bertha Mueller,
    with an introduction by Charles J. Engard. University of Hawaii Press, 1952.

CHAPTER 7. Decarboxylation

The conversion of THCA to THC is discussed in detail in Dussy et al. (2005). The history of
smoking pipes can be found in Sesli and Yeğenoğlu (2018) and in the classic paper by Philips
(1983). Gilman and Zhou (2004) review the history of smoking and survey smoking devices.
Shekhar and Hannah-Shmouni (2020) discuss the dangers of sharing water pipe devices in the
age of COVID-19. Gates et al. (2014) discuss the impact of cannabinoids on the respiratory
tract. Philips (1983) offers a wonderful summary of African smoking habits and devices. The

paper by the "father of Indian pharmacology" and his colleague (Chopra and Chopra 1957) is a remarkable historical pharmacological document on the use of marijuana via foodstuffs in India.

Chopra, I. C., and Ram Nath Chopra. "The Use of the *Cannabis* Drugs in India." *Bulletin on Narcotics* 9, no. 1 (1957): 4–29.

Dussy, F. E., C. Hamberg, M. Luginbühl, T. Schwerzmann, and T. A. Briellmann. "Isolation of Δ9-THCA-A from Hemp and Analytical Aspects Concerning the Determination of Δ9-THC in *Cannabis* Products." *Forensic Science International,* 149, no. 1 (2005): 3–10.

Gates P., A. Jaffe, J. Copeland. "*Cannabis* Smoking and Respiratory Health: Consideration of the Literature." *Respirology* 19, no. 5 (2014): 655–662.

Gilman, Sander L., and Zhou Xun, eds. *Smoke: A Global History of Smoking.* Reaktion Books, 2004.

Philips, John Edward. "African Smoking and Pipes." *Journal of African History* 24, no. 3 (1983): 303–319.

Sesli, Meltem, and E. Dilşat Yeğenoğlu. "Tobacco in the Historical Process." In *Science, Ecology and Engineering Research in the Globalizing World,* edited by Ilia Christov, Eric Strauss, Abd-Alla Gad, and Isa Curebal, 146–152. St. Kliment Ohridski University Press, 2018.

Shekhar, Skand, and Fady Hannah-Shmouni. "Hookah Smoking and COVID-19: Call for Action." *Canadian Medical Association Journal* 192, no. 17 (2020): E462–E462.

CHAPTER 8. THC and CBD in the Body

Grotenhermen (2003), Huestis et al. (1992), and Wall and Perez-Reyes (1981) discuss the metabolism of cannabinoids in the human body. The topical application cutaneous route that CBD can take into the body is discussed by Baswan et al. (2020). Burstein (2015) and Crippa et al. (2018) offer excellent scholarly starting points for understanding CBD and the human body. Helander and Fändriks's (2014) estimation of the surface area of the digestive tract is a lively technical read. For a clear and concise discussion of the structure of skin and its role in topical application, see Hwa et al. (2011).

Baswan, Sudhir M., Allison E. Klosner, Kelly Glynn, Arun Rajgopal, Kausar Malik, Sunghan Yim, and Nathan Stern. "Therapeutic Potential of Cannabidiol (CBD) for Skin Health and Disorders." *Clinical, Cosmetic and Investigational Dermatology* 13 (2020): 927–942.

Burstein, Sumner. "Cannabidiol (CBD) and Its Analogs: A Review of Their Effects on Inflammation." *Bioorganic & Medicinal Chemistry* 23, no. 7 (2015): 1377–1385.

Crippa, José A., Francisco S. Guimarães, Alline C. Campos, and Antonio W. Zuardi.

"Translational Investigation of the Therapeutic Potential of Cannabidiol (CBD): Toward a New Age." *Frontiers in Immunology* 9 (2018): 2009.

Grotenhermen, F. Pharmacokinetics and Pharmacodynamics of Cannabinoids. *Clinical Pharmacokinetics* 42, no. 4 (2003): 327–360.

Helander, Herbert F., and Lars Fändriks. "Surface Area of the Digestive Tract—Revisited." *Scandinavian Journal of Gastroenterology* 49, no. 6 (2014): 681–689.

Huestis, M.A., J. E. Henningfield, and E. J. Cone. "Blood Cannabinoids. I. Absorption of THC and Formation of 11-OH-THC and THC-COOH during and after Smoking Marijuana." *Journal of Analytical Toxicology* 16, no. 5 (1992): 276–282.

Hwa, Charlotte, Eugene A. Bauer, and David E. Cohen. "Skin Biology." *Dermatologic Therapy* 24, no. 5 (2011): 464–470.

Wall, M.E., and M. Perez-Reyes. "The Metabolism of Delta 9-Tetrahydrocannabinol and Related Cannabinoids in Man." *Journal of Clinical Pharmacology* 21, no. S1 (1981): 178S–189S.

Chapter 9. A Plant of (More Than) 1,001 Chemicals

A comprehensive review of the biochemical fecundity of marijuana plants can be found in the following article.

Andre, Christelle M., Jean-François Hausman, and Gea Guerriero. "*Cannabis sativa:* The Plant of the Thousand and One Molecules." *Frontiers in Plant Science* 7 (2016): 19.

The following three papers address the biology of secondary plant compounds in cannabis.

Panche, Archana N., Arvind D. Diwan, and Sadanandavalli R. Chandra. "Flavonoids: An Overview." *Journal of Nutritional Science* 5 (2016): e47.

Popescu-Spineni, Dana Maria, Anca Magdalena Munteanu, Constantin Ionescu-Târgoviște, Costin Militaru, and Alexandru Constantin Moldoveanu. "Cannabis Terpenes in Relation to Human Health." *Revue Roumaine de Chimie* 66, no. 7 (2021): 583–592.

Watts, Sophie, Michel McElroy, Zoë Migicovsky, Hugo Maassen, Robin van Velzen, and Sean Myles. "Cannabis Labelling Is Associated with Genetic Variation in Terpene Synthase Genes." *Nature Plants* 7, no. 10 (2021): 1330–1334.

The entourage effect is addressed in the following citations.

Finlay, David B., Kathleen J. Sircombe, Mhairi Nimick, Callum Jones, and Michelle Glass. "Terpenoids from Cannabis Do Not Mediate an Entourage Effect by Acting at Cannabinoid Receptors." *Frontiers in Pharmacology* 11 (2020): 359.

Russo, Ethan B. "The Case for the Entourage Effect and Conventional Breeding of Clinical Cannabis: No 'Strain,' No Gain." *Frontiers in Plant Science* 9 (2018): 1969.

The structure and biology of trichomes are covered in the following scientific papers.

Conneely, Lee James, Ramil Mauleon, Jos Mieog, Bronwyn J. Barkla, and Tobias Kretzschmar. "Characterization of the *Cannabis sativa* Glandular Trichome Proteome." *PloS One* 16, no. 4 (2021): e0242633.
Livingston, Samuel J., Teagen D. Quilichini, Judith K. Booth, Darren C. J. Wong, Kim H. Rensing, Jessica Laflamme-Yonkman, Simone D. Castellarin, Joerg Bohlmann, Jonathan E. Page, and A. Lacey Samuels. "Cannabis Glandular Trichomes Alter Morphology and Metabolite Content during Flower Maturation." *Plant Journal* 101, no. 1 (2020): 37–56.
Livingston, Samuel J., Kim H. Rensing, Jonathan E. Page, and A. Lacey Samuels. "A Polarized Supercell Produces Specialized Metabolites in Cannabis Trichomes." *Current Biology* 32, no. 18 (2022): 4040–4047.
Serna, Laura, and Cathie Martin. "Trichomes: Different Regulatory Networks Lead to Convergent Structures." *Trends in Plant Science* 11, no. 6 (2006): 274–280.
Tanney, Cailun A. S., Rachel Backer, Anja Geitmann, and Donald L. Smith. "Cannabis Glandular Trichomes: A Cellular Metabolite Factory." *Frontiers in Plant Science* 12 (2021): 1923.
Yeo, Hock Chuan, Vaishnavi Amarr Reddy, Bong-Gyu Mun, Sing Hui Leong, Savitha Dhandapani, Sarojam Rajani, and In-Cheol Jang. "Comparative Transcriptome Analysis Reveals Coordinated Transcriptional Regulation of Central and Secondary Metabolism in the Trichomes of Cannabis Cultivars." *International Journal of Molecular Sciences* 23, no. 15 (2022): 8310.

Ian Tattersall's and my previous books on wine, beer, and spirits and Neil Shubin's excellent book on evolution are referenced below.

Tattersall, Ian, and Rob DeSalle. *A Natural History of Wine.* Yale University Press, 2015.
DeSalle, Rob, and Ian Tattersall. *A Natural History of Beer.* Yale University Press, 2019.
DeSalle, Rob, and Ian Tattersall. *Distilled: A Natural History of Spirits.* Yale University Press, 2022.
Shubin, Neil. *Your Inner Fish: A Journey into the 3.5-Billion-Year History of the Human Body.* Vintage, 2008.

CHAPTER 10. Messy Brains and Marijuana

Many books and articles in the primary literature address the brain and neural structure. I have only included a few.

Cesario, Joseph, David J. Johnson, and Heather L. Eisthen. "Your Brain Is Not an Onion with a Tiny Reptile Inside." *Current Directions in Psychological Science* 29, no. 3 (2020): 255–260.

DeSalle, Rob, and Ian Tattersall. *The Brain: Big Bangs, Behaviors, and Beliefs.* Yale University Press, 2012.

Marcus, Gary. *Kluge: The Haphazard Evolution of the Human Mind.* Houghton Mifflin Harcourt, 2009.

Moroz, Leonid L., and Daria Y. Romanova. "Selective Advantages of Synapses in Evolution." *Frontiers in Cell and Developmental Biology* 9 (2021): 726563.

CHAPTER 11. Brain Science, Bliss, and the Endocannabinoid System

I include two works by the late Dr. Raphael Mechoulam in the listing below. Minelli provides a thorough review of organs and organ systems in animals, while Castillo et al., Lee, Lu and Mackie, and Shao et al. offer helpful summaries of the endocannabinoid system at both the body and molecular levels. Pain and its alleviation by endocannabinoids are discussed in the articles by Mlost et al., Muralidhar et al., and Vučković et al.

Castillo, Pablo E., Thomas J. Younts, Andrés E. Chávez, and Yuki Hashimotodani. "Endocannabinoid Signaling and Synaptic Function." *Neuron* 76, no. 1 (2012): 70–81.

Lee, Martin A. "The Discovery of the Endocannabinoid System." *Prop 215 Era* (2012).

Lu, Hui-Chen, and Ken Mackie. "An Introduction to the Endogenous Cannabinoid System." *Biological Psychiatry* 79, no. 7 (2016): 516–525.

Maccarrone, Mauro. "Missing Pieces to the Endocannabinoid Puzzle." *Trends in Molecular Medicine* 26, no. 3 (2020): 263–272.

Maccarrone, Mauro. "Tribute to Professor Raphael Mechoulam, the Founder of Cannabinoid and Endocannabinoid Research." *Molecules* 27, no. 1 (2022): 323.

Mechoulam, Raphael. "A Delightful Trip along the Pathway of Cannabinoid and Endocannabinoid Chemistry and Pharmacology." *Annual Review of Pharmacology and Toxicology* 63 (2023): 1–13.

Mechoulam, Raphael, and Linda A. Parker. "The Endocannabinoid System and the Brain." *Annual Review of Psychology* 64 (2013): 21–47.

Minelli, Alessandro. "On the Nature of Organs and Organ Systems—A Chapter in the

History and Philosophy of Biology." *Frontiers in Ecology and Evolution* 9 (2021): 745564.

Mlost, Jakub, Marta Bryk, and Katarzyna Starowicz. "Cannabidiol for Pain Treatment: Focus on Pharmacology and Mechanism of Action." *International Journal of Molecular Sciences* 21, no. 22 (2020): 8870.

Muralidhar Reddy, P., Nancy Maurya, and Bharath Kumar Velmurugan. "Medicinal Use of Synthetic Cannabinoids—A Mini Review." *Current Pharmacology Reports* 5 (2019): 1–13.

Shao, Zhenhua, Jie Yin, Karen Chapman, Magdalena Grzemska, Lindsay Clark, Junmei Wang, and Daniel M. Rosenbaum. "High-Resolution Crystal Structure of the Human CB1 Cannabinoid Receptor." *Nature* 540, no. 7634 (2016): 602–606.

Vučković, Sonja, Dragana Srebro, Katarina Savić Vujović, Čedomir Vučetić, and Milica Prostran. "Cannabinoids and Pain: New Insights from Old Molecules." *Frontiers in Pharmacology* 9 (2018): 1259.

Chapter 12. Genes, Genomes, and Cannabis

Kovalchuk et al., Cai et al., and Hurgobin et al. present comprehensive summaries of genomics research in cannabis. Braich et al., Jenkins and Orsburn, Van Bakel et al., and Gao et al. offer the first reports of genome sequence of different cannabis strains. McKernan et al., Laverty et al., and Grassa et al. are the sources used to construct table 12.1. Pisupati et al. discuss the role of transposable elements in the genomic evolution of cannabis.

Braich, Shivraj, Rebecca C. Baillie, German C. Spangenberg, and Noel O. I. Cogan. "A New and Improved Genome Sequence of *Cannabis sativa*." *BioRxiv* (2020): 422592.

Cai, Sen, Zhiyuan Zhang, Suyun Huang, Xu Bai, Ziying Huang, Yiping Jason Zhang, Likun Huang, et al. "CannabisGDB: A Comprehensive Genomic Database for *Cannabis sativa* L." *Plant Biotechnology Journal* 19, no. 5 (2021): 857–859.

Connor, Jason P., Daniel Stjepanović, Bernard Le Foll, Eva Hoch, Alan J. Budney, and Wayne D. Hall. "Cannabis Use and Cannabis Use Disorder." *Nature Reviews Disease Primers* 7, no. 1 (2021): 1–24.

Gao, Shan, Baishi Wang, Shanshan Xie, Xiaoyu Xu, Jin Zhang, Li Pei, Yongyi Yu, Weifei Yang, and Ying Zhang. "A High-Quality Reference Genome of Wild *Cannabis sativa*." *Horticulture Research* 7 (2020).

Grassa, Christopher J., George D. Weiblen, Jonathan P. Wenger, Clemon Dabney, Shane G. Poplawski, S. Timothy Motley, Todd P. Michael, and C. J. Schwartz. "A New Cannabis Genome Assembly Associates Elevated Cannabidiol (CBD) with Hemp Introgressed into Marijuana." *New Phytologist* 230, no. 4 (2021): 1665–1679.

Hurgobin, Bhavna, Muluneh Tamiru-Oli, Matthew T. Welling, Monika S. Doblin, Antony

Bacic, James Whelan, and Mathew G. Lewsey. "Recent Advances in *Cannabis sativa* Genomics Research." *New Phytologist* 230, no. 1 (2021): 73–89.

Jenkins, Conor, and Ben Orsburn. "The First Publicly Available Annotated Genome for Cannabis Plants." *BioRxiv* (2019): 786186.

Kovalchuk, I., M. Pellino, P. Rigault, R. Van Velzen, J. Ebersbach, J. R. Ashnest, M. Mau, et al. "The Genomics of Cannabis and Its Close Relatives." *Annual Review of Plant Biology* 71 (2020): 713–739.

Laverty, Kaitlin U., Jake M. Stout, Mitchell J. Sullivan, Hardik Shah, Navdeep Gill, Larry Holbrook, Gintaras Deikus, et al. "A Physical and Genetic Map of *Cannabis sativa* Identifies Extensive Rearrangements at the THC/CBD Acid Synthase Loci." *Genome Research* 29, no. 1 (2019): 146–156.

McKernan, Kevin J., Yvonne Helbert, Liam T. Kane, Heather Ebling, Lei Zhang, Biao Liu, Zachary Eaton, et al. "Sequence and Annotation of 42 Cannabis Genomes Reveals Extensive Copy Number Variation in Cannabinoid Synthesis and Pathogen Resistance Genes." *BioRxiv* (2020): 894428.

McPartland, John M., Isabel Matias, Vincenzo Di Marzo, and Michelle Glass. "Evolutionary Origins of the Endocannabinoid System." *Gene* 370 (2006): 64–74.

Pisupati, Rahul, Daniela Vergara, and Nolan C. Kane. "Diversity and Evolution of the Repetitive Genomic Content in *Cannabis sativa*." *BMC Genomics* 19, no. 1 (2018): 1–9.

Van Bakel, Harm, Jake M. Stout, Atina G. Cote, Carling M. Tallon, Andrew G. Sharpe, Timothy R. Hughes, and Jonathan E. Page. "The Draft Genome and Transcriptome of *Cannabis sativa*." *Genome Biology* 12, no. 10 (2011): 1–18.

CHAPTER 13. Putting the Cannabis Genome to Work

The URLs for the websites referred to in this chapter are as follows.

Allbud	https://www.allbud.com/
Cannabis.info	https://www.cannabis.info/en/
CannaSOS (CSOS)	https://cannasos.com/
Grow Marijuana	https://grow-marijuana.com/
ILGM	https://www.ilovegrowingmarijuana.com/
Leafly (Lly)	https://www.leafly.com/
NCSM	https://www.ncsm.nl/english/
Wikileaf	https://www.wikileaf.com/

Aardema and DeSalle discuss the utility of public cannabis databases, while Jikomes and Zoorob comment on the validity of THC and CBD concentration estimates in different strains of cannabis. Dolgin provides a cogent review of the biochemical production potential of THCA and CBDA using biomanufacturing. Ren et al. discuss the recent successes of *CRISPR* technology on grape cultivation, while Simiyu et al. and Hesami et al. review the success genetic researchers have had with cannabis. The rest of the papers listed here examine the associations of cannabis traits with genetic variants.

Aardema, Matthew L., and Rob DeSalle. "Can Public Online Databases Serve as a Source of Phenotypic Information for Cannabis Genetic Association Studies?" *PloS One* 16, no. 2 (2021): e0247607.

Andre, Christelle M., Jean-François Hausman, and Gea Guerriero. "*Cannabis sativa:* The Plant of the Thousand and One Molecules." *Frontiers in Plant Science* 7 (2016): 19.

Chen, Xuan, Hong-Yan Guo, Qing-Ying Zhang, Lu Wang, Rong Guo, Yi-Xun Zhan, Pin Lv, et al. "Whole-Genome Resequencing of Wild and Cultivated Cannabis Reveals the Genetic Structure and Adaptive Selection of Important Traits." *BMC Plant Biology* 22, no. 1 (2022): 1–16.

Conneely, Lee James, Ramil Mauleon, Jos Mieog, Bronwyn J. Barkla, and Tobias Kretzschmar. "Characterization of the *Cannabis sativa* Glandular Trichome Proteome." *PloS One* 16, no. 4 (2021): e0242633.

Dolgin, Elie. "The Bioengineering of Cannabis." *Nature* 572, no. 7771 (2019): S5–S5.

Finlay, David B., Kathleen J. Sircombe, Mhairi Nimick, Callum Jones, and Michelle Glass. "Terpenoids from Cannabis Do Not Mediate an Entourage Effect by Acting at Cannabinoid Receptors." *Frontiers in Pharmacology* 11 (2020): 359.

Grassa, Christopher J., George D. Weiblen, Jonathan P. Wenger, Clemon Dabney, Shane G. Poplawski, S. Timothy Motley, Todd P. Michael, and C. J. Schwartz. "A New Cannabis Genome Assembly Associates Elevated Cannabidiol (CBD) with Hemp Introgressed into Marijuana." *New Phytologist* 230, no. 4 (2021): 1665–1679.

Halpin-McCormick, Anna, Karolina Heyduk, Michael Kantar, Nick Batora, Rishi Masalia, Kerin Law, and Eleanor Kuntz. "Phylogenetic Resolution of the Cannabis Genus Reveals Extensive Admixture." *BioRxiv* (2022): 499013.

Hesami, Mohsen, Austin Baiton, Milad Alizadeh, Marco Pepe, Davoud Torkamaneh, and Andrew Maxwell Phineas Jones. "Advances and Perspectives in Tissue Culture and Genetic Engineering of Cannabis." *International Journal of Molecular Sciences* 22 (2021): 5671.

Jikomes, Nick, and Michael Zoorob. "The Cannabinoid Content of Legal Cannabis in Washington State Varies Systematically across Testing Facilities and Popular Consumer Products." *Scientific Reports* 8, no. 1 (2018): 1–15.

Jin, Dan, Philippe Henry, Jacqueline Shan, and Jie Chen. "Classification of Cannabis Strains in the Canadian Market with Discriminant Analysis of Principal Components Using Genome-Wide Single Nucleotide Polymorphisms." *PloS One* 16, no. 6 (2021): e0253387.

Kovalchuk, I., M. Pellino, P. Rigault, R. Van Velzen, J. Ebersbach, J. R. Ashnest, M. Mau, et al. "The Genomics of Cannabis and Its Close Relatives." *Annual Review of Plant Biology* 71 (2020): 713–739.

Livingston, Samuel J., Teagen D. Quilichini, Judith K. Booth, Darren C. J. Wong, Kim H. Rensing, Jessica Laflamme-Yonkman, Simone D. Castellarin, Joerg Bohlmann, Jonathan E. Page, and A. Lacey Samuels. "Cannabis Glandular Trichomes Alter Morphology and Metabolite Content during Flower Maturation." *Plant Journal* 101, no. 1 (2020): 37–56.

Livingston, Samuel J., Kim H. Rensing, Jonathan E. Page, and A. Lacey Samuels. "A Polarized Supercell Produces Specialized Metabolites in Cannabis Trichomes." *Current Biology* 32, no. 18 (2022): 4040–4047.

Panche, Archana N., Arvind D. Diwan, and Sadanandavalli R. Chandra. "Flavonoids: An Overview." *Journal of Nutritional Science* 5 (2016): e47.

Petit, Jordi, Elma M. J. Salentijn, Maria-João Paulo, Christel Denneboom, and Luisa M. Trindade. "Genetic Architecture of Flowering Time and Sex Determination in Hemp (*Cannabis sativa* L.): A Genome-Wide Association Study." *Frontiers in Plant Science* 11 (2020): 569958.

Popescu-Spineni, Dana Maria, Anca Magdalena Munteanu, Constantin Ionescu-Târgovişte, Costin Militaru, and Alexandru Constantin Moldoveanu. "Cannabis Terpenes in Relation to Human Health." *Revue Roumaine de Chimie* 66, no. 7 (2021): 583–592.

Prentout, Djivan, Olga Razumova, Bénédicte Rhoné, Hélène Badouin, Hélène Henri, Cong Feng, Jos Käfer, Gennady Karlov, and Gabriel A. B. Marais. "An Efficient RNA-Seq-Based Segregation Analysis Identifies the Sex Chromosomes of *Cannabis sativa*." *Genome Research* 30, no. 2 (2020): 164–172.

Ren, Chong, Yanping Lin, and Zhenchang Liang. "CRISPR/Cas Genome Editing in Grapevine: Recent Advances, Challenges and Future Prospects." *Fruit Research* 2, no. 1 (2022): 1–9.

Ren, Guangpeng, Xu Zhang, Ying Li, Kate Ridout, Martha L. Serrano-Serrano, Yongzhi Yang, Ai Liu, et al. "Large-Scale Whole-Genome Resequencing Unravels the Domestication History of *Cannabis sativa*." *Science Advances* 7, no. 29 (2021): eabg2286.

Russo, Ethan B. "The Case for the Entourage Effect and Conventional Breeding of Clinical Cannabis: No 'Strain,' No Gain." *Frontiers in Plant Science* 9 (2018): 1969.

Simiyu, David Charles, Jin Hoon Jang, and Ok Ran Lee. "Understanding *Cannabis sativa* L.: Current Status of Propagation, Use, Legalization, and Haploid-Inducer-Mediated Genetic Engineering." *Plants* 11, no. 9 (2022): 1236.

Sirangelo, Tiziana M., Richard A. Ludlow, and Natasha D. Spadafora. "Multi-Omics Approaches to Study Molecular Mechanisms in *Cannabis sativa.*" *Plants* 11, no. 16 (2022): 2182.

Watts, Sophie, Michel McElroy, Zoë Migicovsky, Hugo Maassen, Robin van Velzen, and Sean Myles. "*Cannabis* Labelling Is Associated with Genetic Variation in Terpene Synthase Genes." *Nature Plants* 7, no. 10 (2021): 1330–1334.

Welling, Matthew T., Lei Liu, Tobias Kretzschmar, Ramil Mauleon, Omid Ansari, and Graham J. King. "An Extreme-Phenotype Genome-Wide Association Study Identifies Candidate Cannabinoid Pathway Genes in Cannabis." *Scientific Reports* 10, no. 1 (2020): 18643.

Woods, Patrick, Brian J. Campbell, Timothy J. Nicodemus, Edgar B. Cahoon, Jack L. Mullen, and John K. McKay. "Quantitative Trait Loci Controlling Agronomic and Biochemical Traits in *Cannabis sativa.*" *Genetics* 219, no. 2 (2021): iyab099.

Woods, Patrick, Nicholas Price, Paul Matthews, and John K. McKay. "Genome-Wide Polymorphism and Genic Selection in Feral and Domesticated Lineages of *Cannabis sativa.*" *G3 Genes| Genomes| Genetics* 13, no. 2 (2023): jkac209.

Yeo, Hock Chuan, Vaishnavi Amarr Reddy, Bong-Gyu Mun, Sing Hui Leong, Savitha Dhandapani, Sarojam Rajani, and In-Cheol Jang. "Comparative Transcriptome Analysis Reveals Coordinated Transcriptional Regulation of Central and Secondary Metabolism in the Trichomes of Cannabis Cultivars." *International Journal of Molecular Sciences* 23, no. 15 (2022): 8310.

Zoorob, Michael J. "The Frequency Distribution of Reported THC Concentrations of Legal Cannabis Flower Products Increases Discontinuously around the 20% THC Threshold in Nevada and Washington State." *Journal of Cannabis Research* 3, 6 (2021), https://doi.org/10.1186/s42238-021-00064-2.

CHAPTER 14. Modern Medicinal Cannabis

The websites listed here are helpful in understanding the basics of clinical trials. The Fauci reference concerns the COVID-19 vaccine clinical trials process. The Asaad et al. and van Dorn papers examine the impact of COVID-19 on clinical trials. Witik, Mlost et al., Scheau et al., Vučković et al., and Gras et al., address the issue of cannabis treatment of pain. Billakota et al. and Devinsky et al. are papers on epilepsy and cannabis. The report by Powell concerns synthetic marijuana and its dangers. Harrer et al., Namdar et al., and Smolinski et al. focus on cannabis in clinical trials.

https://www.clinicaltrials.gov

https://www.fda.gov/patients/clinical-trials-what-patients-need-know/basics-about-clinical-trials

https://www.nia.nih.gov/health/what-are-clinical-trials-and-studies
https://clinicaltrialsexplained.com/

Asaad, Malke, Nilofer Khan Habibullah, and Charles E. Butler. "The Impact of COVID-19 on Clinical Trials." *Annals of Surgery* 272, no. 3 (2020): e222.

Billakota, Santoshi, Orrin Devinsky, and Eric Marsh. "Cannabinoid Therapy in Epilepsy." *Current Opinion in Neurology* 32, no. 2 (2019): 220–226.

Devinsky, Orrin, J. Helen Cross, Linda Laux, Eric Marsh, Ian Miller, Rima Nabbout, Ingrid E. Scheffer, Elizabeth A. Thiele, and Stephen Wright. "Trial of Cannabidiol for Drug-Resistant Seizures in the Dravet Syndrome." *New England Journal of Medicine* 376, no. 21 (2017): 2011–2020.

Fauci, Anthony S. "The Story behind COVID-19 Vaccines." *Science* 372, no. 6538 (2021): 109.

Gras, Airy, Oriane Hidalgo, Ugo D'Ambrosio, Montse Parada, Teresa Garnatje, and Joan Vallès. "The Role of Botanical Families in Medicinal Ethnobotany: A Phylogenetic Perspective." *Plants* 10, no. 1 (2021): 163.

Harrer, Stefan, Pratik Shah, Bhavna Antony, and Jianying Hu. "Artificial Intelligence for Clinical Trial Design." *Trends in Pharmacological Sciences* 40, no. 8 (2019): 577–591.

Mlost, Jakub, Marta Bryk, and Katarzyna Starowicz. "Cannabidiol for Pain Treatment: Focus on Pharmacology and Mechanism of Action." *International Journal of Molecular Sciences* 21, no. 22 (2020): 8870.

Namdar, Dvora, Omer Anis, Patrick Poulin, and Hinanit Koltai. "Chronological Review and Rational and Future Prospects of Cannabis-Based Drug Development." *Molecules* 25, no. 20 (2020): 4821.

Powell, Devin. "The Spice of Death: The Science behind Tainted Synthetic Marijuana." *Scientific American,* April 17, 2018.

Riboulet-Zemouli, Kenzi. "'Cannabis' Ontologies I: Conceptual Issues with Cannabis and Cannabinoids Terminology." *Drug Science, Policy and Law* 6 (2020): 1–37.

Scheau, Cristian, Ioana Anca Badarau, Livia-Gratiela Mihai, Andreea-Elena Scheau, Daniel Octavian Costache, Carolina Constantin, Daniela Calina, Constantin Caruntu, Raluca Simona Costache, and Ana Caruntu. "Cannabinoids in the Pathophysiology of Skin Inflammation." *Molecules* 25, no. 3 (2020): 652.

Smolinski, Nicole E., Ruba Sajdeya, Robert Cook, Yan Wang, Almut G. Winterstein, and Amie Goodin. "Proceedings of the 2022 Cannabis Clinical Outcomes Research Conference." *Medical Cannabis and Cannabinoids* 5, no. 1 (2022): 138–141.

Van Dorn, Aaron. "COVID-19 and Readjusting Clinical Trials." *Lancet* 396, no. 10250 (2020): 523–524.

Vučković, Sonja, Dragana Srebro, Katarina Savić Vujović, Čedomir Vučetić, and Milica

Prostran. "Cannabinoids and Pain: New Insights from Old Molecules." *Frontiers in Pharmacology* 9 (2018): 1259.

Witek Jr., Theodore J. "Please Don't Call It Medical Marijuana Unless It Is; But It Probably Isn't." *Canadian Journal of Public Health* 112, no. 1 (2021): 74–77.

CHAPTER 15. Legalize It?

The following papers concern the legalization of cannabis and the history of legislation concerning cannabis.

Collins, John. "A Brief History of Cannabis and the Drug Conventions." *American Journal of International Law* 114 (2020): 279–284.

Cox, Chelsea. "The Canadian Cannabis Act Legalizes and Regulates Recreational Cannabis Use in 2018." *Health Policy* 122, no. 3 (2018): 205–209.

Kamin, Sam. "Marijuana Law Reform in 2020 and Beyond: Where We Are and Where We're Going." *Seattle University Law Review* 43 (2020): 883.

Patton, David V. "A History of United States Cannabis Law." *Journal of Law and Health* 34 (2020): 1–29.

Rotermann, Michelle. "Looking Back from 2020, How Cannabis Use and Related Behaviours Changed in Canada." *Statistics Canada,* April 21, 2021, https://www.doi.org/10.25318/82-003-x202100400001-eng.

CHAPTER 16. Dangerous?

Johnson et al. and Connor et al. address cannabis use disorder. The classic book by Grinspoon and Bakalar remains helpful. The Beaulieu paper reviews the toxicity of cannabis in animals. Nutt and colleagues report on the dangers of several drugs, and the Rickner et al. paper describes dabbing and its dangers. Payne et al. and Dixit et al. discuss the impact of cannabis use on testicular function. Memedovich et al. and Pisanti and Bifulco consider the health dangers of marijuana use, and DeSalle and Tattersall focus on the health and addiction issues of alcohol.

Beaulieu, Pierre. "Toxic Effects of Cannabis and Cannabinoids: Animal Data." *Pain Research and Management* 10, suppl. A (2005): 23a–26a.

Connor, Jason P., Daniel Stjepanović, Bernard Le Foll, Eva Hoch, Alan J. Budney, and Wayne D. Hall. "Cannabis Use and Cannabis Use Disorder." *Nature Reviews Disease Primers* 7, no. 1 (2021): 16.

DeSalle, Rob, and Ian Tattersall. *A Natural History of Beer.* Yale University Press, 2019.

Dixit, V. P., Chandra Lekha Gupta, and Meera Agrawal. "Testicular Degeneration and Necrosis Induced by Chronic Administration of Cannabis Extract in Dogs." *Endokrinologie* 69, no. 3 (1977): 299–305.

Dixit, V. P., V. N. Sharma, and N. K. Lohiya. "The Effect of Chronically Administered Cannabis Extract on the Testicular Function of Mice." *European Journal of Pharmacology* 26, no. 1 (1974): 111–114.

Grinspoon, Lester, and James B. Bakalar. *Marihuana, the Forbidden Medicine.* Yale University Press, 1997.

Johnson, Emma C., Ditte Demontis, Thorgeir E. Thorgeirsson, Raymond K. Walters, Renato Polimanti, Alexander S. Hatoum, Sandra Sanchez-Roige, et al. "A Large-Scale Genome-Wide Association Study Meta-Analysis of Cannabis Use Disorder." *Lancet Psychiatry* 7, no. 12 (2020): 1032–1045.

Memedovich, K. Ally, Laura E. Dowsett, Eldon Spackman, Tom Noseworthy, and Fiona Clement. "The Adverse Health Effects and Harms Related to Marijuana Use: An Overview Review." *Canadian Medical Association Open Access Journal* 6, no. 3 (2018): E339–E346.

Nutt, David J., Leslie A. King, and Lawrence D. Phillips. "Drug Harms in the UK: A Multicriteria Decision Analysis." *Lancet* 376, no. 9752 (2010): 1558–1565.

Payne, Kelly S., Daniel J. Mazur, James M. Hotaling, and Alexander W. Pastuszak. "Cannabis and Male Fertility: A Systematic Review." *Journal of Urology* 202, no. 4 (2019): 674–681.

Pisanti, Simona, and Maurizio Bifulco. "Medical Cannabis: A Plurimillennial History of an Evergreen." *Journal of Cellular Physiology* 234, no. 6 (2019): 8342–8351.

Rickner, Shannon S., Dazhe Cao, Kurt Kleinschmidt, and Steven Fleming. "A Little 'Dab' Will Do Ya' in: A Case Report of Neuro- and Cardiotoxicity Following Use of Cannabis Concentrates." *Clinical Toxicology* 55, no. 9 (2017): 1011–1013.

CHAPTER 17. Contemporary Cannabis Culture in Nine Graphs

The New Cannabis Ventures link is https://www.newcannabisventures.com/cannabis-stock-index/. Several other websites listed here were used to compile the figures in this chapter. Knapp et al. examines the use of cannabis in teens. Wyse and Luria provide an excellent summary of the state of the art for patents up until 2021. ElSohly et al. and Chandra et al. consider the potency issue of cannabis. Dills et al. describe the legal status of marijuana in various countries and states. O'Grady details patterns of spending by the US federal government on cannabis research.

https://www.newcannabisventures.com/cannabis-stock-index/
https://pharmaboardroom.com/facts/global-medical-cannabis-regulatory-landscape/
https://www.grandviewresearch.com/industry-analysis/legal-marijuana-market
https://only420.com/9462/images/countries-that-support-cannabis-legalization/
https://www.vettafi.com/indexing/index/canabizp
https://www.portfoliowealthglobal.com/cannabis-legalization-the-dam-has-burst/

https://www.vice.com/en/article/jgp8ny/threats-to-children-parents-posting-social
-media-photos-on-instagram
https://vividmaps.com/cannabis-legality/#google_vignette
https://cfah.org/marijuana-statistics/
https://inhalemd.com/blog/who-funds-marijuana-research-united-states/
https://drugabusestatistics.org/drug-related-crime-statistics/

Blaszczak-Boxe, Agata. "Potent Pot: Marijuana Is Stronger Now Than It Was 20 Years Ago." *Live Science,* February 8, 2016. https://www.livescience.com/53644 -marijuana-is-stronger-now-than-20-years-ago.html.

Chandra, Suman, Mohamed M. Radwan, Chandrani G. Majumdar, James C. Church, Tom P. Freeman, and Mahmoud A. ElSohly. "New Trends in Cannabis Potency in USA and Europe during the Last Decade (2008–2017)." *European Archives of Psychiatry and Clinical Neuroscience* 269, no. 1 (2019): 5–15.

Dills, Angela K., Sietse Goffard, Jeffrey Miron, and Erin Partin. "The Effect of State Marijuana Legalizations: 2021 Update." Cato Institute, *Policy Analysis* 908 (2021): 1–37.

ElSohly, Mahmoud A., Suman Chandra, Mohammed Radwan, Chandrani Gon Majumdar, and James C. Church. "A Comprehensive Review of Cannabis Potency in the United States in the Last Decade." *Biological Psychiatry: Cognitive Neuroscience and Neuroimaging* 6, no. 6 (2021): 603–606.

Kay, Kip. *Marijuana in the Movies: The Complete Guide to the Hollywood High.* Media-Green Press, 1999.

Knapp, Ashley A., Dustin C. Lee, Jacob T. Borodovsky, Samantha G. Auty, Joy Gabrielli, and Alan J. Budney. "Emerging Trends in Cannabis Administration among Adolescent Cannabis Users." *Journal of Adolescent Health* 64, no. 4 (2019): 487–493.

O'Grady, Cathleen. "Cannabis Research Data Reveals a Focus on Harms of the Drug." *Science* 369 (2020): 1155.

Wyse, Joseph, and Gilad Luria. "Trends in Intellectual Property Rights Protection for Medical Cannabis and Related Products." *Journal of Cannabis Research* 3 (2021): 1–22.

Index

Page numbers in *italics* refer to figures; numbers in **bold** refer to tables.